P9-CKM-010

WILD WILD FLOWERS
OF THE WEST

by

EDITH S. KINUCAN
and
PENNEY R. BRONS

Published By
Kinucan & Brons
Box 765
Ketchum, Idaho 83340

Copyright © 1985
by
Edith S. Kinucan and Penney R. Brons

All rights reserved.   This book or parts thereof may not be reproduced
in any form without permission in writing from the Publisher, except
that a reviewer may quote brief passages in a review.

First Edition 1977
Second Edition 1979
Third Edition 1985

Library of Congress Catalog Card Number: 85-90969
ISBN 0-9615444-0-6

INTRODUCTION

If you are an individual who is interested in the minute scientific details needed to key plants out to species, subspecies, varieties, etc., this book is not for you. This is a volume for the non-scientist who is interested in becoming familiar with wildflowers - their identity, uses and interesting facts - without having to delve into glossaries, diagrams and scientific terminology. Hopefully, you will be able to recognize many wildflowers by simply comparing them to the pictures in this book.

I have yet to meet a person who didn't react with pleasure and even awe to the sight of a field of brilliant wildflowers or to the delicate beauty of just one small blossom. Nor have I met a man, woman or child who didn't display a degree of curiosity about plants in general. Such curiosity is probably an innate characteristic of human nature, since from his very beginning, mankind has been totally dependent on plants for existence.

Wildflowers are intriguing in many ways. Not only are they wonderful to look at, but they are also indicators of the composition and quality of the environment in which they live. Each individual wildflower must live in a habitat that meets its particular requirements. Some flowers can tolerate a wide range of environmental conditions. That is, they can stand great heat or cold, dryness or wetness, rocky, poor soils or rich, organic soils. Others are very sensitive, and must have optimum temperatures, an exact amount of moisture, specific minerals in the soil, and other specific conditions.

More can be learned about the health and vigor of a particular environment from wildflowers than from dozens of scientific testing instruments. For example, a beautiful field of lupine is, sadly, often an indication that the land has been abused by overgrazing, usually by domestic livestock such as sheep. A flamboyant display of white wyethia tells the informed person that an area has probably been mismanaged. An abundance of mullein shows that the land has been disturbed either by a natural disaster or by human intervention by road building, bulldozing, or other activity. Fireweed growing in profusion is often an indication that a fire has passed through the area. Dying vegetation may be the result of insect invasion, drought, air pollution or other disrupting environmental factors, or it may be a natural process of plant succession. A little investigation can produce a great deal of information. Healthy thriving plants usually mean a healthy environment. But, always, the kinds and conditions of wildflowers tell a story of more than beauty to the informed individual.

I have always felt that to know the wildflowers is to feel at home anywhere that I have lived. Being familiar with wildflowers can be not only comforting, but might also mean the difference between survival or death. It is always good to know which flowers you can depend upon for food should you become lost or stranded. Too, it helps to be aware that not all plants are beautiful, nutritious and fragrant. Some are deadly poisonous (water hemlock), cause skin irritations with blisters or rashes (stinging nettle) and even smell bad (skunkweed). Knowledge of these facts can be important to your health and well being.

Wildflowers can fool you, too. Sometimes what you think is a flower is really something else (Indian paintbrush), and what you consider to be a petal of a flower is actually the entire flower (members of the sunflower family). In some instances wildflowers that are considered weeds are a potential source of food. After all, as Ralph Waldo Emerson said, "And what is a weed? A plant whose virtues have not been discovered."

Penney and I offer you this book with the hope that you will find it as rewarding and pleasurable as it has been for us during its creation. For your convenience, we have arranged the plants by color and have grouped them according to families in each color section. Among the plants included in this book are at least 65 edible species, 35 medicinal ones and 25 that are poisonous. Welcome to the third edition of *Wild Wildflowers of the West.*

E.S.K.

ACKNOWLEDGEMENTS

We are most grateful to Dr. Karl Holte of Idaho State University for his help with plant identification. Dr. Fred D. Johnson of the University of Idaho provided valuable criticism and suggestions for which we thank him. We appreciate the help that Ricky Bosted gave in making the original indexing both fun and easy. Many thanks to AnMarie Giddings who cheerfully helped with the revised index and with proofreading. The patience and expertise of Paul Bosted helped produce the beautiful flower prints of the first edition and we offer them as a memorial to a special friend who is sorely missed. We also thank Ken Kinucan and John Brons for their patience and support, and Robert Kinucan for providing the picture of pasque flower. We would be remiss in not expressing our thanks to those many friends who have given us encouragement and support in our efforts. We are especially grateful to Idaho's Governor John Evans, Dr. R. J. Vogl, Dr. Brian Capon, and Professor Emeritus Roland C. Ross of California State University at Los Angeles, Dr. Marcus W. Jordin, University of Arkansas for Medical Science, Poison Control Group, and Dr. William J. Keppler, Dean of the School of Arts and Sciences, Boise State University for their generous words of praise.

E.S.K. & P.R.B.

DEDICATION

To Ken and John with love and gratitude.

QUEEN'S CUP. *Clintonia uniflora* (Lily family: Liliaceae)
SIZE: 3 to 8 in. tall with two or three basal leaves; single flower about 1 in. broad.
FOUND: Wet to moist soil, partial shade on forest floor under conifers from British Columbia to central Sierra Nevada.
BLOOMS: Late May to July.
EDIBLE: Leaves.
USES AND FACTS: The fruit of *Clintonia* is a blue berry that looks good, but is not edible except by birds and other animals. Ruffed grouse like the berries. An eastern species, *C. borealis,* has edible leaves (Ellis and Dykeman, 1982). Leaves should be gathered when very young and can be used for salad or boiled about 10 minutes and seasoned. Cowlitz Indians applied juice to cuts or sore eyes when lids stuck together upon awakening (Gunther, 1981). Because the plant has underground stems, there are usually several clusters of leaves in one area. The single flower inspired the species name *uniflora,* meaning one flower. **Caution:** leaves are similar in appearance to potentially poisonous plants.   This plant should be left alone unless needed in an emergency because of its small size and beauty.

1

FALSE HELLEBORE. *Veratrum californicum* (Lily family: Liliaceae)
SIZE: 3 to 8 ft. high. Flowers 1/2 in. broad.
FOUND: Wet meadows and open areas of valleys in mountains to about 9,000 ft. throughout the West.
BLOOMS: June to early August.
DANGER: Poisonous.
USES AND FACTS: False hellebore is poisonous to livestock, deer, elk and human beings. The seeds are poisonous to chickens and probably to other birds as well. It can be fatal if taken in large quantities as it contains several alkaloids such as veratrin which cause the following symptoms: watering of the mouth, vomiting, diarrhea, stomach pains, general paralysis and spasms. Severe cases may result in shallow breathing, slow pulse, lower temperatures, convulsions and death (Hardin & Arena, 1974). The dry powdered roots have been used as a garden insecticide. Indians reportedly used raw roots, crushed and mashed, to apply to snake bite wounds on humans and other animals. A decoction of the root was taken as a tea for venereal disease. Spaniards made a poison from fermented juices of the roots which they used for arrow heads (Sweet, 1962). The alkaloids in false hellebore are used medicinally to slow the heartbeat and to lower blood pressure.

2

FALSE SOLOMONSEAL   **WILD LILY OF THE VALLEY**

FALSE SOLOMONSEAL. *Similicina racemosa.* WILD LILY OF
THE VALLEY. *S. stellata* (Lily family: Lilaceae)
SIZE: 1 to 2 ft. tall. Flowers small.
FOUND: Moist soil of open and shaded areas, especially along
streams in valleys and mountains up to 9,000 ft. from British
Columbia to Labrador, south to Virginia, Texas and California, and in
most Western states.
BLOOMS: May to July.
EDIBLE: Shoots and leaves.
USES AND FACTS: Craighead, *et al.*, (1963) say that young shoots
and leaves can be used as a green. Elk eat shoots and leaves. Hardin
and Arena (1974) warn that *S. racemosa* (false Solomon's seal)
berries can be eaten raw, but with caution.   Kirk (1970) says the
starchy aromatic rootstocks of *S. racemosa* and related species may be
eaten.   They should be soaked overnight in lye to remove the
bitterness, then parboiled to remove the lye. Rootstocks make a good
pickle. Young shoots can be eaten as a potherb. Berries of all species
are edible, but they are laxative if eaten in large quantities. Cooking
removes the purgative element and makes the berries taste better.
Berries are usually red or greenish-red with small purple spots.

3

DEATH CAMAS. *Zigadenus venenosus* (Lily family: Liliaceae)
MOUNTAIN DEATH CAMAS. *Zigadenus elegans*

SIZE: 1 to 2 ft. tall. Flowers of *Z. elegans* about 3/8 in. long. Those of *Z. venenosus* smaller.

FOUND: Species of *Zigadenus* are found in dry rocky areas as well as in moist soils of meadows up to about 6,000 ft.; *Z. elegans* in moist soils of meadows, along streams and springs from 6,000 to 12,000 ft. Another species, *Z. paniculatus*, occurs in drier foothills and plains.

BLOOMS: June to August, depending on elevation.

DANGER: Poisonous.

USES AND FACTS: The alkaloids, such as zygadenine, are concentrated mainly in the bulb and cause muscular weakness, slow heartbeat, subnormal temperature, stomach upset with pain, vomiting, diarrhea and excessive watering of the mouth. Children have been poisoned by both bulb and flowers (Hardin & Arena, 1974). Death camas sometimes grows with blue camas (*Camassia* sp.) which is edible. The bulbs look very similar, and accidental poisoning sometimes occurred among Indians when bulbs were collected when the plants were not in bloom. Balls (1962) states that Indians used crushed bulbs as poultices for boils, bruises, strains, rheumatism, and some places, for rattlesnake bites. Occasionally the bulbs were roasted and applied warm.

4

DEATH CAMAS

MOUNTAIN
DEATH CAMAS

SEGO LILY. *Calochortus nutallii* (Lily family: Liliaceae)
SIZE: 8 to 20 in. high. Flowers 2 to 3 in. across.
FOUND: On well-drained dry to moist ground of plains, hillsides, open woods and sagebrush communities from 5,000 to 9,000 ft. in Western states.
BLOOMS: June to July.
EDIBLE: Bulbs.
USES AND FACTS: There are 40 species of *Calochortus* throughout the West. The bulbous root is sweet and nutritious, and was used by the Indians. It can be eaten raw or cooked. The flavor is supposed to be improved by slowly steaming masses of them in fire pits, or by roasting them over a smokey fire (Kirk, 1970). When boiled it tastes like potato. The bulbs may be dried and ground into flour. *C. nutalli* is the state flower of Utah. Mormon pioneers used this plant for food. *Cal* is Greek for beautiful and *chortus* means green herbage. Other species of this genus are also edible. Tubers are eaten by bears and rodents; seed pods by domestic sheep and bighorn sheep (Craighead, *et al.,* 1963).

6

WAKE ROBIN; BIRTHROOT.   *Trillium ovatum* (Lily family: Liliaceae)
SIZE: 8 to 16 in. tall.   Single flower 1 to 2 in. across with three conspicuous white petals that turn pink, then rose-colored with age.
FOUND: Usually in woods still moist from snow melt, or boggy areas in partial shade from low valleys to 7,000 ft. from British Columbia to California.
BLOOMS: Late March to June, when snow disappears.
EDIBLE: Leaves.
DANGER: Roots are a powerful emetic (induce vomiting).
USES AND FACTS: The root of *Trillium* was used by some Indians as a presumed aid to childbirth (Kirk, 1970). Craighead, *et al.,* 1963, also refer to this plant as having been used during childbirth. It doesn't really sound too appealing - an emetic while in labor! The Makah Indians had a better idea. They pounded the bulb and rubbed it on the body as a love medicine (Gunther, 1981). Some Indians used the juice of the boiled bulbs as an eye wash. The leaves may be boiled and eaten as greens. However, when the leaves of the plant are removed, the bulb dies from lack of nourishment, so please restrict the use of this plant to emergency situations if it is used at all.

BEAR GRASS. *Xerophyllum tenax* (Lily family: Lliaceae)
SIZE: Flowering stalk up to 3 ft. tall. Tough, sharp, grass-like leaves form a tussock at the base of the stalk.
FOUND: Mountain slopes, open woods and alpine meadows from British Columbia to California in Western states.
BLOOMS: June to September, depending on elevation.
EDIBLE: Roots.
USES AND FACTS: The fibrous root can be eaten roasted or boiled. Leaves were used by Indians for overlay or decoration on baskets and in cloth making (Gunther, 1981). This species does not bloom every year - perhaps once every 5 to 7 years. Fields of blooming bear grass look very spectacular. *Xero* is the Greek word for dry, *phyllum* is Greek for plant, and *tenax* is Latin for holding fast, or tough.

WHITE BOG ORCHID. *Habenaria dilatata* (Orchid family: Orchidaceae)
SIZE: 1 to 2 ft. tall. Flowers small on a spike.
FOUND: In wet soils of swamps and bogs, along springs and streams from lowest valleys to 10,000 ft. in most Western states.
BLOOMS: May to early August, depending on elevation.
EDIBLE: Tubers.
USES AND FACTS: Tubers are radish-like and can be eaten raw or cooked. However, orchids are quite rare, and should be used as food only in an emergency. All species of orchid have edible tubers (Kirk, 1970).

GRASS OF PARNASSUS. *Parnassia fimbriata* (Saxifrage family: Saxifragaceae)

SIZE: 6 to 12 in. tall. Flower 3/4 in. across.

FOUND: Around springs, streambanks, boggy areas, generally in shade, from about 5,000 ft. in mountains to timberline in Western states.

BLOOMS: July to August.

USES AND FACTS: The name *Parnassus* is taken from a mountain in Greece sacred to Apollo. It was the home of the Muses, the nine goddesses of song and poetry. *Fimbriata* is Latin for fringed and refers to the fringed petals.

CANADIAN DOGWOOD; BUNCHBERRY. *Cornus canadensis*
(Dogwood family: Cornaceae)
SIZE: Low, creeping subshrub 2 to 8 in. tall. Flowers small, incon-
spicuous, surrounded by four showy white bracts which give the appear-
ance of a four-petaled white flower.
FOUND: In moist areas of coniferous forests from Alaska to Cali-
fornia and New Mexico from sea level to 8,000 ft.
BLOOMS: May to July.
EDIBLE: Fruit.
USES AND FACTS: Fruits may be eaten raw or cooked (Turner,
1975). Dried bark or root that had been boiled was used as a mild
stimulant or to treat cold and fever. The fresh bark is cathartic (laxative).
Feathered bark was used as a toothbrush by old timers (Clarke, 1977).
Leaves of the related species, *C. stolonifera,* red osier dogwood, a large
shrub, were used as a substitute for smoking tobacco. Flowering dogwood, *C. nuttallii.* is the state flower of both Virginia and North Caro-
lina.

11

PEARLY EVERLASTING.   *Anaphalis margaritacea* (Sunflower family: Compositae)
SIZE: 1 to 3 ft. tall. Flowers about 1/4 in., in dense clustered heads.
FOUND: Dry to moist soil in foothills and mountains to treeline. Species of *Anaphalis* occur in Canada and much of the U.S.
BLOOMS: Late June to late August.
USES AND FACTS: This plant makes a good winter bouquet as it dries like a strawflower when picked soon after blooming. *Margaritacea* means pearly (from the Greek, *margarites*, a pearl).

WHITE WYETHIA or **WHITE MULE-EARS.** *Wyethia helianthoides*
(Sunflower family: Compositae)
SIZE: 8 to 20 in. tall. Flower heads 2 to 5 in. broad.
FOUND: Moist to wet soil of meadows, open woods and seepage
areas from foothills to 8,000 ft. in Northwestern states.
BLOOMS: May to early July, depending on elevation.
EDIBLE: Roots.
USES AND FACTS: Indians fermented roots by heating them on
stones one or two days. The flavor is agreeable and sweet. Roots
were also used as a poultice for relief of pains and bruises. A
decoction of leaves was used as a bath and produced profuse
sweating. NEVER TAKE DECOCTION INTERNALLY -
POISONOUS (Sweet, 1962). Flowers and young leaves are eaten by
livestock, deer and elk. The plants grow in large patches and seem to
increase rapidly in overgrazed areas.

PUSSYTOES or **EVERLASTING.** *Antennaria* sp. (Sunflower family: Compositae)
SIZE: 2 to 12 in. tall. Flowers small, 1/8 in. across, in clusters.
FOUND: Dry to moist soil of prairies, valleys and mountain sides to about 9,000 ft. throughout the West.
BLOOMS: Late May to early August.
USES AND FACTS: The gum of the stalks can be chewed, and is claimed to be nourishing and pleasant tasting. It was used by some Western Indians. Many species of *Antennaria* produce seed without fertilization. Plants are usually male or female, and in some species, male plants are rare or unknown. When picked soon after blooming, flowers can be kept as dry winter bouquets and are amenable to dye (Craighead, *et al.*, 1963).

14

FERNLEAF FLEABANE; CUTLEAF DAISY. *Erigeron compositus*
(Sunflower family: Compositae)
SIZE: 1 to 10 in. tall. Flower heads 1/2 to 1 in. across.
FOUND: Dry, rocky or sandy soil from foothills to above timberline
throughout the West.
BLOOMS: May to July.
USES AND FACTS: The fleabanes vary in palatability to game and
livestock, but are generally poor forage. The cutleaf daisy increases
with overgrazing and is used as an indicator of range abuse. It
becomes especially abundant on over-used cattle ranges (Craighead,
et al., 1963). Fleabanes or daisies are often confused with aster. Both
belong to the aster tribe. The ray flowers (those that resemble petals)
of the fleabane are narrow, fine and numerous, whereas those of the
aster are generally wider and less numerous.

15

YARROW. *Achillea millefolium* (Sunflower family: Compositae)
SIZE: 1 to 4 ft. tall. Flowers small, in heads. The plants are taller at lower elevations and shorter towards treeline.
FOUND: Frequently seen along roadsides. Occurs in dry to moist soil from low valleys to above timberline throughout North America.
BLOOMS: June through August, depending on elevation.
DANGER: Poisonous.
USES AND FACTS: *A. millefolium* is a naturalized species from Eurasia, and is thought to contain an alkaloid posion. It closely resembles the native yarrow, *A. lanulosa,* which is not poisonous, and which can be dried, boiled in water, strained and administered to remedy a run-down condition or disordered digestion. This brew also makes a nourishing broth (Kirk, 1970). Indians used yarrow for a number of disorders: to stop bleeding and heal inflamation; juice in the eye to remove redness; oil made from the plant to stop falling hair (Sweet, 1962). Achilles reportedly used a similar plant to treat his warriors' wounds (hence the name *Achillea*). Yarrow has a pungent odor. It gives milk a disagreeable flavor when consumed by cows. Since the two species look so much alike, we suggest that you avoid consumption of yarrow.

16

DWARF BUCKWHEAT. *Eriogonum ovalifolium* (Buckwheat family: Polygonaceae)
SIZE: Less than 10 in. tall. Flowers small, in clusters.
FOUND: Dry slopes and flats from 5,000 to 7,000 ft. from British Columbia to California throughout the West.
BLOOMS: May to July.
USES AND FACTS: There are about 150 species of buckwheat in North America, mostly in the West. They range in color from white, yellow, to orange and reddish, and in size from less than 10 in. to several feet in height. Some are important as "bee-plants," and the plants are good food for domestic sheep. Seeds remain available during the winter for wildlife. This is a good "winter bouquet" plant as it dries well and retains its color. (See yellow flowers.) Sweet (1962) states that Indians used a decoction of leaves for headache and stomach pains; also a tea from flowers for eyewash, high blood pressure and bronchial ailments. Stems and leaves were boiled for tea to treat bladder trouble.

AMERICAN BISTORT.   *Polygonum bistortoides* (Buckwheat family:
Polygonaceae)
SIZE: 8 to 28 in. tall.   Flowers small, in clustered head.
FOUND: From valley floors to above timberline in wet meadows,
along streams, in mountains, canyons and among rock debris at high
elevations throughout Western states.
BLOOMS: Early June to August, depending on elevation.
EDIBLE: Roots, young leaves.
USES AND FACTS: Rootstocks are starchy and slightly astringent
when eaten raw.   When boiled, they are sweeter; best roasted on
coals.   Young greens can be used for salad or potherb (Elias & Dykeman,
1982). Approximately 45 related species are found over most of the
U.S.

18

CRYPTANTHA.   *Cryptantha torreyana* (Borage family: Boraginaceae)
SIZE: 4 to 26 in. tall. Flowers small, in clusters.
FOUND: Open or half-shaded slopes from 1,500 to 7,500 ft. in elevation in Western states.
BLOOMS: May to August.
USES AND FACTS: There are 65 species of *Cryptantha* found in the western part of North America. *Cryptos* is Greek for hidden, and *anthos* for flower. The plant was called *Cryptantha* because of the minute flowers of the first species named. Most species of *Cryptantha* are small (less than 18 in. high), coarsely hairy or bristly plants with rather inconspicuous flowers (Taylor & Valum, 1974). Most species are white; a few are yellow. Flowers resemble the forget-me-not.

ELKWEED; GREEN GENTIAN. *Frasera speciosa* (Gentian family: Gentianaceae)

SIZE: 2 to 5 feet tall. Flowers 1/2 to 3/4 inches on long stalks.

FOUND: In open areas with medium to dry to moist soil in mountains from about 6,500 ft. to over 10,000 ft. throughout Western states from Montana to California and New Mexico.

BLOOMS: June into August.

EDIBLE: Root.

USES AND FACTS: The fleshy root may be eaten raw, boiled or roasted and is good when mixed with potherbs or raw greens (Kirk, 1970). Many members of the Gentian family have been used medicinally throughout the world. *F. carolinensis* has been used as an emetic and cathartic, but *F. speciosa* will not produce such effects unless eaten in excessive quantities. Elkweed is eaten by elk and cattle, especially in spring when the basal leaves are young and tender. Elkweed is biennial. The first year growth produces a cluster of long strap-like leaves. Flowering occurs the second year. Both stages are shown in picture above.

STICKSEED; FALSE FORGET-ME-NOT: *Hackelea patens* (Borage family: Boraginaceae)
SIZE: 1 to 2 ft. tall.   Flowers about 1/3 in. broad.
FOUND: Foothills to about 8,000 ft. in moist to medium dry soil from British Columbia and in Western states to California.
BLOOMS: June to early August.
USES AND FACTS: The flowers of this plant closely resemble those of dwarf forget-me-not *(Eritrichium)*.   The small seeds are nutlets that have prickles with fish hook-like barbs on the end.   They cling tenaciously to animal fur and clothing so are widely disseminated by this means.   If you walk through a stand of stickseed during the fruiting time, you will certainly know it!

EVENING PRIMROSE. *Oenothera caespitosa* (Evening primrose family: Onagraceae)

SIZE: Leaves clustered around rootstock close to ground 2 to 6 in. long.   Flowers 2 to 4 in. broad.

FOUND: Plains and valleys to about 8,000 ft. on dry soil, stony slopes, sandy places and ridges from Saskatchewan to Washington and south to California and New Mexico over much of Western U.S.

BLOOMS: Late May into July.

EDIBLE: Leaves and roots.

USES AND FACTS: Leaves and roots are edible but bitter tasting (Winegar, 1982).   There are about 200 species of *Oenothera*, mainly in North and South America.   *Caespitosa* is Latin, meaning growing in tufts.   *Oeno* is a Greek root for wine, and *thera* is Greek for hunting.   However, Hitchcock and Cronquist (1978) say that Oenothera is a Greek name used by Theophrastus, said to mean wine-scented.   The white flowers of this species turn pink or red with age.   The flowers open at night, but contrary to popular opinion, also stay open during the day.

CINQUEFOIL. *Potentilla glandulosa* (Rose family: Rosaceae)
SIZE: 1 to 2-1/2 ft. tall. Flowers 1/2 to 3/4 in. broad.
FOUND: Dryish to moist soil in open places from plains and hills to 8,000 ft. elevation in mountains throughout the West.
BLOOMS: May to July, depending on elevation.
USES AND FACTS: There is no data available concerning the edibility of this species of cinquefoil. There are about 16 species of cinquefoil in Idaho. One of these, *P. anserina* (silverweed), a low growing plant spreading by means of runners, has a yellow flower and resembles *P. glandulosa* somewhat. The roots of silverweed are edible raw or cooked. Leaves of bush cinquefoil can be used for tea. (See yellow flowers.)

23

WILD STRAWBERRY. *Fragaria virginiana* (Rose family: Rosacea)
SIZE: Low plant with runners. Flowers about 3/4 in. broad.
FOUND: Moist soils of woods, meadows and along streams from low valleys to treeline. Species of strawberry are found throughout temperate North America.
BLOOMS: Early May throughout summer.
EDIBLE: Fruit and leaves (in beverage).
USES AND FACTS: The berries are small but sweet, and can be used in the same manner as are domestic strawberries. Some Indians made a tea-like beverage from the leaves. The berries are used by many birds and mammals - so much so, that they are often hard to find. The strawberry derived its name from a practice of laying straw around the cultivated plants to keep the fruits from becoming soiled in wet weather. The berries are eaten by ruffed grouse, robins, turtles, small rodents, black and grizzly bears as well as many other wildlife species (Craighead, *et al.*, 1963).

MEADOW-RUE. *Thalictrum   fendleri* (Buttercup family: Ranunculaceae)

SIZE: 1 to 5 ft. tall. Flowers small.

FOUND: Moist soil near streams and meadows, often in shade and thickets, occasionally on dry slopes, from 4,000 to 10,000 ft. in coniferous forests throughout the West.

BLOOMS: May to August.

USES AND FACTS: There is no data available concerning the edibility or poisonousness of this plant. The Greek name *thalictron* means meadow-rue. The plant was named by Dioscorides, a Greek physician who lived from 41 to 68 A.D. He held an exalted position in botany and medicine until the 17th Century and named nearly 600 plants and almost 1,000 drugs in *De Materia Medica*. Meadow-rue plants are unisexual. That is, the male flower is found on one plant, and the female flower on another.

FEMALE FLOWER

MALE FLOWER

25

VIRGIN'S BOWER. *Clematis ligusticifolia* (Buttercup family: Ranunculaceae)

SIZE: Semi-woody vine. Flowers small, about 1/3 in. across, but occur so profusely that they impart a white color to the whole mass of growth. The seeds, (which are most conspicuous) are adorned with long feathery tails.

FOUND: Grows over bushes and trees along streams in streamside woodlands, and along roadsides and ditches from Canada to California throughout the West.

BLOOMS: May to August, with extended flowering as the vine continues to grow.

DANGER: Poisonous. The entire plant is toxic, containing protoanemonin which has a direct irritant and vesicant effect on skin and mucous membranes. Upon swallowing, there is intense pain and inflamation of the mouth, often with blistering and ulceration. Saliva is profuse, bloody vomiting and diarrhea occur with accompanying abdominal cramps (Lampe and McCann, 1985).

USES AND FACTS: Sweet (1962) says that Spanish Americans called virgin's bower Yerba de Chivato (herb of the goat). They used a decoction from this plant to wash wounds. Indians used the white portion of the bark to treat fever, leaves and bark for shampoo, and a decoction of leaves on horses for sores and cuts. Fibers were used for snares and carrying nets. *Pharmocopia* says it is useful for the treatment of skin diseases, ulcers, colds and many eruptions. In the 16th century it was reportedly used internally and in powder form to cure bone pains. According to Craighead, *et al.,* (1963), a decoction was used by Indians for colds and sore throats. Hunters with cold feet can use the fuzz from the seeds to insulate their boots.

BANEBERRY. *Actaea arguta* (Buttercup family: Ranunculaceae)
SIZE: 1 to 3 ft. tall. Flowers small, clustered.
FOUND: In moist or wet places, often in shaded areas, along streams, springs and in boggy areas from valleys to 9,000 ft. in mountains throughout the West.
BLOOMS: May to early July. Fruit remains until mid-August.
DANGER: Poisonous.
USES AND FACTS: The berries are bright red or white. All parts, but mostly roots and berries, contain a poisonous glycoside or essential oil which causes acute stomach cramps, headache, increased pulse, vomiting, delirium, dizziness, and circulatory failure. As few as six berries can cause severe symptoms that can persist for hours (Hardin & Arena, 1974). The berries are very attractive and are particularly dangerous to children!

GOLD THREADS. *Coptis occidentalis* (Buttercup family: Ranunculaceae)

SIZE: Plant up to 10 in. high. Flowers small, white and often missed. Seedpods are most conspicuous.

FOUND: Moist woods and rocky places in foothills and mountains from British Columbia to California.

BLOOMS: Early spring, March when the snow is gone. Seeds form in early spring. You have to be quick and observant to catch this one in bloom! Spot the evergreen leaves when snow melts, then watch for the flowers. It's worth it!

DANGER: Poisonous.

USES AND FACTS: This plant belongs to the buttercup family and probably contains the same toxic substance that other members of the family do, so watch that children don't eat the attractive seedpods. *C. laciniata* and *C. asplenifolia* are related species. *Coptis* is from the Greek *koptis* meaning to cut, referring to the dissected leaves.

28

WATER CRESS.   *Rorippa nasturtium-aquaticum* (Mustard family: Cruciferae)

SIZE: 4 to 24 in. high, stems usually prostate.   Flowers small.

FOUND: Common in quiet water or on wet banks of streams below 8,000 ft. throughout the West.

BLOOMS: June through August.

EDIBLE: Leaves.

DANGER: Many streams carry a parasitic protozoan called *Giardia lamblia* which is spread by domestic and wild animals.   It causes giardiasis, an intestinal disorder, which is miserable and difficult to cure, so eat wild water cress at your own risk.   You can disinfect it by washing in water that contains iodide purifying tablets.

USES AND FACTS: This plant is native to the Old World and has become naturalized here.   It is excellent raw in salads or can be cooked in various ways as a potherb or used as seasoning in soups and meat dishes.   It can also be used with rice or in white sauce.   (Harrington, 1967).

BITTERROOT. *Lewisia rediviva* (Purslane family: Portulacaceae)
SIZE: 1 to 3 in. high. Flowers 1 to 2 in. broad.
FOUND: Usually in rocky, dry soil of valleys or foothills; stony slopes, ridges and mountain summits to about 8,000 ft. from British Columbia to California throughout the West.
BLOOMS: Late April through June into July, depending on elevation.
EDIBLE: Roots.
USES AND FACTS: Roots were used in quantity by Indians who usually dug them before the flower developed. The outer root cover peels off easily, leaving a white fleshy core which is bitter until boiled or baked. It also can be powdered to form a meal. The root is jelly-like in appearance when boiled. Reservation Indians still use it. *Rediviva* refers to the plants ability to return to life after the root has been dried for weeks or months. Bitterroot is the state flower of Montana. The flower color ranges from white to dark pink.

From the most forbidding of environments, there can spring forth strength and beauty.

STINGING NETTLE. *Urtica dioica* (Nettle family: Urticaceae)
SIZE: 1-1/2 to 6-1/2 ft. tall. Flowers small and inconspicuous.
FOUND: Along streams, ditches and wastelands and in woods; frequently in disturbed areas throughout North America below 9,000 ft.
BLOOMS: June to September.
EDIBLE: Shoots.
DANGER: Skin irritant.
USES AND FACTS: Nettles have stiff stinging (urticating) hairs which have a fluid containing formic acid that is very irritating to the skin. Formic acid is injected by bees and ants, too, when they sting. Some people have a much more allergic reaction than do others. Nettles are edible. Cooking removes the stinging hairs. The most common use is as a potherb. The tender shoots, 6 to 8 in. tall, are best, including the pink underground stem. Nettle is high in vitamin C. It can be used as a tea and also as a substitute for rennet to coagulate milk. Beer and wine may be made from nettles when added to dandelion flowers, lemon juice, ginger root, brown sugar and yeast. The fibers of mature stems can be retted out and spun into cloth (used in Europe). A yellow dye can be made from boiled roots.

LOCOWEED. *Astragalus* sp. (Pea family: Leguminosae)
SIZE: Species range in size from 3-8 in. up to 3 ft. in height. Flowers resemble sweetpeas and vary in size and color (white, pink, purple, yellow).
FOUND: Species are found from plains and hills to rocky ridges of mountains in dry soils throughout the West.
BLOOMS: Late May to August, depending on elevation.
DANGER: Poisonous.
USES AND FACTS: There are approximately 52 species of *Astragalus* in Idaho; 400 in North America; 2,000 in the world. All should be considered dangerous as they contain an alkaloid poisonous to animals. Some species absorb selenium from the soil which produces the "loco disease" in livestock. Sweet (1962) says that Indians chewed the plants to cure sore throat and to reduce swellings. The boiled root was made into a decoction to wash granulated eyelids and for toothaches.

WOODLAND-STAR. *Lithophragma parviflora* (Saxifrage family: Saxifragaceae)
SIZE: 8 to 20 in. tall. Flowers 1/4 in. to 1/2 in. long.
FOUND: Rich, medium-dry soil from low valleys to about 9,000 ft. in mountains from British Columbia to California in Western states.
BLOOMS: April through June.
USES AND FACTS: *Lithos* is Greek for stone or rock, and *phragma* is Greek for hedge. *Parviflora* is Latin for small flower. The bulblets of this plant are eaten by rodents, chukars, and probably Hungarian partridges (Craighead, *et al.,* 1963). It is not known whether or not humans can eat them. When in doubt, don't!

VIOLET.   *Viola beckwithii* (Violet family: Violaceae)
SIZE: 2 to 4 in. high.   Flowers about 3/4 in. long.
FOUND: Dry, gravelly places, often among shrubs from 3,000 to 6,000 ft. from Oregon to California throughout the West.
BLOOMS: May to June.
EDIBLE: Leaves and flowers.
USES AND FACTS: All species of *Viola* are edible.   The young leaves and flower buds can be used in salads.   Leaves and flowers can also be used as potherb, and the leaves can be used for tea.   The flowers can be candied.   (See yellow.)   Four states, Illinois, New Jersey, Rhode Island and Wisconsin have some species of violets as their state flower.

FALSE STRAWBERRY. *Hesperochiron pumilus* (Waterleaf family: Hydrophyllaceae)

SIZE: About 2 in. high. Flowers about 5/8 in. broad.

FOUND: Moist flats and meadows from 1,200 to 9,000 ft. in the West from Washington to California.

BLOOMS: April to July, depending on elevation.

USES AND FACTS: The name *hesperos* (Greek) means western, and *chiron* means a centaur skilled in medicine. This might indicate that the plant had medicinal uses in the past. However, we found no uses listed. *Pumil* is Latin for diminutive. The flowering time for this plant is short, so watch closely for it in early spring; later at higher elevations.

WATER HEMLOCK. *Cicuta* sp. (Carrot or parsley family: Umbelliferae)

SIZE: 2 to 7 ft. tall. Flowers small, in clusters.

FOUND: Species of *Cicuta* occur in moist habitats in thickets, in meadows, along streams, around spring heads, in marshes, seepage areas and roadside ditches from low valleys to about 8,000 ft. in mountains throughout the U.S.

BLOOMS: Mid-June to late August.

DANGER: Poisonous.

USES AND FACTS: The poisonous chemicals are found mostly in the rootstock and much less in above-ground parts. The root is extremely poisonous; one mouthful is sufficient to kill an adult. It is so often mistaken for wild parsnip or artichoke, that deaths are frequent. Children have been poisoned by making peashooters and whistles from hollow stems. Symptoms: diarrhea, violent convulsions and spasms, tremors, extreme stomach pain, dilated pupils, frothing at mouth, delirium and death (Hardin & Arena, 1974). Many members of the carrot or parsley family are edible. However, no member of this family should be eaten until positively identified, and then only sparingly until proven safe. WHEN IN DOUBT, DON'T! Socrates probably drank tea from a near relative of water hemlock. and he didn't live to tell about it. Watch for purple stripes or spots on the stem to help identify this plant.

COW PARSNIP. *Heracleum lanatum* (Carrot or parsley family: Umbelliferae)

SIZE: 3 to 8 ft. tall. Flowers small and in clusters. Leaves from 4 to 12 in. broad.

FOUND: Rich, damp soil of prairies and mountains, especially along streams, roadsides and in open woods, occurring in most of temperate North America from sea level to about 8,500 ft.

BLOOMS: Late May to early July.

EDIBLE - DANGER: Many species similar in appearance to this plant are deadly poisonous. Be very cautious! Also, the juice and hairs of the outer skin can cause facial blisters. If you are sure of the plant, you can eat the roots and stems.

USES AND FACTS: This plant was used for food by Indians and Eskimos, and formerly had wide use as a medicine. The young stems are sweet and succulent and can be peeled and eaten raw (Craighead, *et al.*, 1963). Harrington (1968) says the plant smells and tastes rank. The cooked roots taste like rutabaga, but, Standley (1943) suggests that they may be poisonous. (You have to be pretty courageous to eat this plant!) The hollow basal portions of the plant can be cut into short lengths, dried and used as a salt substitute. Leaves may be dried and burned and the ashes used as a salt substitute (Kirk, 1970). Tender young shoots can be used as potherb or in salad. Older stems can be peeled, and the tender inner tissue eaten raw or cooked.

VALERIAN. *Valeriana* sp. (Valerian family: Valerianaceae)
SIZE: 1 to 3 ft. tall. Flowers small, in clusters.
FOUND: Moist to wet soils in open areas of woods. Species of *Valeriana* occur from hills to almost timberline throughout most Western states.
BLOOMS: May to early August, depending on elevation.
EDIBLE: Roots, seeds.
USES AND FACTS: The roots of some species of *Valeriana* have an unpleasant smell and taste, but become quite palatable when steamed in a stone-lined fire pit or steamed by modern methods for about 24 hours. They can then be eaten as they are or made into soup, or dried and ground into a flour and made into bread. Seeds may be eaten raw, but are best parched. These plants were used as food by a number of Western Indians (Kirk, 1970). Species of valerian are cultivated as drug plants as they contain valerian which is used as a mild stimulant, an antispasmodic and for nervous disorders.

PHLOX. *Phlox* sp. (Phlox family: Polemoniaceae)
SIZE: 4/5 to 4 in. high, usually in mats. Flowers about 3/4 in. across.
FOUND: On dry rocky, gravelly and medium-moist soils in areas from 4,000 to 8,000 ft. in most plant communities from the desert to the mountains throughout the West.
BLOOMS: May through July, depending on elevation.
USES AND FACTS: There are about 45 species of wild phlox in North America, all resembling domestic phlox which was developed from the wild phlox. Species range in color from white to pink to lavender. This attractive plant is often found in areas that have been over-grazed by sheep. The leaves are quite prickly and unpalatable so are probably ignored by most grazing animals.

GLACIER LILY; DOGTOOTHED VIOLET. *Erythronium grandiflorum* (Lily family: Liliaceae)
SIZE: 3 to 16 in. tall. Flowers large and showy.
FOUND: Sagebrush slopes and valleys up to 12,000 ft. in the mountains. Moist soil along streams in mountains, shaded woods and meadows from British Columbia to Colorado.
BLOOMS: April, as soon as snow recedes and blooms about a month.
EDIBLE: Bulbs, leaves, seedpods.
DANGER: Hart & Moore (1976) warn that leaves and bulbs sometimes impart burning sensation and vomiting if eaten in quanitity.
USES AND FACTS: Root can be boiled or dried for winter storage. Green seedpods taste like string beans when boiled. Leaves and fresh green seedpods can be used as greens (Kirk, 1970). Bears like the bulbs and seedpods are eaten by deer, elk, and other animals. (Craighead, *et al.*, 1963).

MULE-EARS.   *Wyethia amplexicaulis* (Sunflower family: Compositae).
SIZE: 1 to 2 ft. tall. Flower heads 2 to 3 in. broad.
FOUND: Moderately dry to moist soil on open hillsides in higher valleys and mountains to about 7,500 ft. throughout western North America.
BLOOMS: May to July.
USES AND FACTS: One species of *Wyethia* was used for treating poison oak (*Rhus diversiloba*) by making a strong extract from the root and applying it to affected area (Balls, 1962). *Amplexicaulis* means stem-clasping. Mule-ears often hybridizes with *W. helianthoides* to produce a pale yellow flower. *Wyethia* is not good for livestock feed and is often considered a range pest. Indians used the root as food (*see* white wyethia).

ARROWLEAF BALSAMROOT. *Balsamorhiza sagittata* (Sunflower family: Compositae).
SIZE: 8 to 16 in. high. Flower heads 2 to 4 in. across.
FOUND: Dry soils of valleys and hills and in mountains up to 8,000 ft. throughout the West.
BLOOMS: Late April to July.
EDIBLE: All parts.
USES AND FACTS: The large root can be eaten raw or cooked (boiled, roasted, etc.). In spring, young leaves and stems can be used in salad or boiled as greens. The leaves and stems become tough and fibrous as they get older, but are still edible. Seeds are excellent roasted and may be ground into a nutritious flour. When a cup of the flour is added to the recipe for a loaf of white bread, the result is delicious (Kirk, 1970). Sweet (1962) lists several medicinal uses by Indians, e.g., root decoction for rheumatism or headache; dry powdered root or mashed root applied as dressing for syphlitic sores and for swellings or insect bites. The gummy root sap was swallowed for consumption. Balsamroot often grows in large patches, covering entire mountains at some elevations.

SUNFLOWER. *Helianthus annuus* (Sunflower family: Compositae)
SIZE: 1 to 8 ft. tall. Flower head 3 to 5 in. across.
FOUND: Dry to medium-moist soil in open areas, waste places, abandoned fields and roadsides throughout the U.S. to about 7,000 ft. in the mountains.
BLOOMS: July to September.
EDIBLE: Seeds.
USES AND FACTS: The sunflower is the state flower of Kansas. Some cultivated sunflowers reach a height of 20 feet. The seeds of the wild sunflower are as edible as those of the cultivated plants. Oil from sunflowers is high grade and is used in cooking, in margarine and in paints. Indians also obtained fiber from the stems and a yellow dye from the flowers, as well as oil from the seeds. Jerusalem artichoke (*H. tuberosus*) is a species of sunflower with an edible root.

DANDELION. *Taraxacum officinale* (Sunflower family:
Compositae.
SIZE: 2 to 20 in. high. Flower head about 1 in. across.
FOUND: Almost everywhere in moist to wet soil of fields, thickets,
lawns, open woods from sea level to high mountains throughout the
West.
BLOOMS: Early May until late fall.
EDIBLE: Leaves, roots, flowers and seeds.
USES AND FACTS: The large fleshy root is an official drug, and has
been used for centuries as a tonic, diuretic and mild laxative. The
tender young leaves can be used as a potherb; roots can be used in
salads; flowers make a delicately flavored wine; seeds can be eaten
raw for emergency food. The plant is supposed to be high in vitamins
A and C. The flowers and leaves are a favorite spring and summer
food of Canada geese, ruffed grouse, and are used by elk, deer, black
and grizzly bear and porcupines (Craighead, *et al.,* 1963). Clarke
(1977) says that the roasted root has been used as a substitute for
coffee.

45

FALSE DANDELION. *Agoseris* sp. (Sunflower family: Compositae)

SIZE: 4 to 25 in. tall. Flower heads about 1 in. broad.

FOUND: Moderately dry to moist, or even wet soils of meadows, roadsides and open areas in mountains to most elevations throughout the West.

BLOOMS: May into August, depending on elevation.

USES AND FACTS: Like the true dandelion, this plant has a milky sap. The solidified juice of *Agoseris* was reportedly chewed by western Indians. It is moderately grazed by livestock. Domestic sheep are very fond of it. About a dozen species of *Agoseris* occur in the Rocky Mountains, two of which are pictured here.

HEART-LEAFED ARNICA.   *Arnica cordifolia* (Sunflower family: Compositae)
SIZE: 8 to 14 in. tall.   Flower heads 2 in. across.
FOUND: Moist soil, usually in open woods under stands of quaking aspen, ponderosa and lodgepole pine from foothills to 9,000 ft. throughout the West.
BLOOMS: May to August.
DANGER: Poisonous.
USES AND FACTS: *A. cordifolia* is an official drug plant.   All parts of this plant may be used as a drug, but the flower is the most potent.   When given orally or intravenously, the drug raises the body temperature.   As a salve used externally, it aids in keeping down infection (Craighead, *et al.,* 1963).   *Cordifolia* means heart-leafed.   *A. montana,* a native of Europe which is occasionally cultivated in rock gardens of the U.S. and Canada has an extract used in medicine.   The flowers and roots of this species have caused vomiting, drowsiness and coma when consumed by children (Hardin & Arena, 1974).

47

WESTERN GROUNDSEL.

Senecio integerrimus (Sunflower family: Compositae)

SIZE: 8 to 30 in. tall. Flower heads 1/4 to 1/2 in. across and in clusters.

FOUND: Medium-dry to moist soil from prairies to near timberline throughout Western states.

BLOOMS: May to early July and August, depending on elevation.

DANGER: Might be poisonous.

USES AND FACTS: There are over 1,000 species of *Senecio* distributed over the earth, about 40 of which occur in the Rocky Mountains. The genus *Senecio* probably contains more species than any other in the plant kingdom. Groundsel contains alkaloids poisonous to cattle and horses, but is not often consumed in quantity (Craighead, *et al.,* 1963). We do not recommend it for human consumption.

48

GOLDENROD. *Solidago* sp. (Sunflower family: Compositae)
SIZE: 1 to 6 ft. tall. Flowers small, numerous, in heads.
FOUND: Moist soil along fences, highways, open waste places and open woods up to 8,000 ft. from British Columbia to California in the West.
BLOOMS: Late July to September.
EDIBLE: Leaves of *S. missouriensis.*
USES AND FACTS: There are about 100 species of goldenrod, mostly native to North America, but a few occur in South America and Eurasia. There ae 12 species in the Rocky Mountains, eight in Idaho. Sweet (1962) states that Indians boiled leaves and used the decoction to wash wounds and ulcers, then sprinkled powdered leaves on the wounds. The same remedy was used for saddle sores on horses. Spanish Americans used the fresh plant mixed with soap for a plaster to bind sore throats. *S. missouriensis*, found along streams and in open pine forests throughout the West (Kirk, 1970) has leaves that make good potherbs, or tea (from dried leaves and flowers). Sap from goldenrod has a high rubber content, and efforts have been made in the past to breed and cultivate the plants for a domestic rubber supply.
A species of goldenrod is the state flower of both Nebraska and Kentucky.

WOOLY YELLOW DAISY. *Eriophyllum lanatum* (Sunflower family: Compositae)

SIZE: 4 to 24 in. tall. Flower heads 1/4 to 3/4 in. broad.

FOUND: Dry soil in open areas of foothills and mountains to about 8,000 ft. from British Columbia to California in the West.

BLOOMS: May to July.

USES AND FACTS: This plant is covered with wooly hairs which help to prevent evaporation of water from the leaves. Many plants have this adaptation which allows them to survive in very dry locations. *Lanatum* means wooly.

GOATSBEARD; SALSIFY. *Tragopogon dubius* (Sunflower family: Compositae)

SIZE: 1 to 4 ft. high. Flower heads 1 to 1-1/2 in. broad.

FOUND: Medium-dry to moist soil along roads, fences and disturbed areas from lowest elevations to about 7,000 ft. throughout the West.

BLOOMS: June through August.

EDIBLE: Roots and young stems.

USES AND FACTS: The fleshy roots may be eaten raw or cooked. When just a few inches high, the young stems with the bases of the lower leaves may be used as potherbs. The coagulated milky juice was used for chewing by various Indians. It was considered a remedy for indigestion. In ancient Greece, Italy and other Old World areas, linen pads were soaked in the distilled juice and applied to bleeding sores and wounds. Pliny recorded that the juice, when mixed with human milk, was a cure-all for disorders of the eyes. (Kirk, 1970). Salsify was introduced from Europe. There are three species in the Rocky Mountains. The seed heads can be used for dried flower arrangements when gently sprayed with hairspray, plastic or colored enamels.

51

SKUNK CABBAGE. *Lysichitum americanum* (Cala lily family: Araceae)

SIZE: Leaves grow from root stalk in a cluster 1 to 5 ft. long and 4 to 16 in. broad.   Flowers are small and sunken into a fleshy axis (spadix) that is partially enclosed in a showy yellow bract (spathe). The spadix is 8 to 12 in. long.

FOUND: Swampy ground, marshes and wet woods, rarely in dense shade in Western states from Alaska to California.

BLOOMS: As soon as snow melts in early spring.

EDIBLE: Root, white part of underground stalk, spadix.

DANGER: Long sharp crystals of calcium oxalate found in this plant.

USES AND FACTS: The calcium oxalate cyrstals can become embedded in the mucous membranes and cause intense irritation and burning. Prolonged cooking and storage will eliminate the crystals (Turner, 1975).   The roots should never be eaten raw.   They can be boiled, but were never highly prized as food by Indians.   The stalk was roasted on hot rocks, and the spadix was buried and a fire built on top or steamed.   The flowering part is very strong and can cause sickness if too many are consumed.   Makah Indian women reportedly chewed the raw root and drank the liquid to clean the bladder.   Others did the same thing to purify the blood.   Leaves were used as a poultice on cuts and swellings.   The leaves were also used as "waxed paper" by Indians for lining baskets, berry drying rocks and steaming pits, according to Gunther (1981).

YELLOW PAINTBRUSH. *Castilleja sulphurea* (Figwort family: Scrophulariaceae)
SIZE: 4 to 16 in. high. Flowers greenish, up to 1 in. long, but concealed by colored bracts that are mistaken for the flowers.
FOUND: Moist to medium-dry soil of plains, foothills and up toward timberline in mountains, from Montana to Idaho, and south to Nevada and Colorado.
BLOOMS: May to July, depending on elevation.
EDIBLE: Flowers.
USES AND FACTS: There are 19 species of paintbrush in Idaho, one of which, *C. linariaefolia,* is the state flower of Wyoming. Kirk (1970) says that many, perhaps all species have flowers that may be eaten raw. However, when selenium is present in the soil, many species absorb it and become toxic. Therefore, care should be taken to eat only normal quantities, if it is eaten at all. Paintbrush is a semi-parasitic plant. That is, it grows with its roots close to those of another plant and derives some of its nutrition from the neighboring plant. Positive identification of various species is very difficult as they look so much alike. Species range in color from white, yellow to varying shades of pink, red, and orange.

53

BUTTER AND EGGS. *Linaria vulgaris* (Figwort family: Scrophulariaceae)

SIZE: 8 in. to 2 ft. tall. Flowers 1/2 to 1-1/2 in. long.

FOUND: Along highways and hillsides throughout West, up to about 5,000 ft.

BLOOMS: July to September.

USES AND FACTS: This plant is a native of Europe that has escaped cultivation in the U.S., and is considered a weed. It is similar to the cultivated snapdragons and is very beautiful. It is worth planting in your garden or yard, but must be kept under control.

YELLOW MONKEYFLOWER. *Mimulus guttatus* (Figwort family: Scrophulariaceae)

SIZE: 2 in. to 2 ft. tall. Flowers 1/2 to 1-1/2 in. long. Ranges in size from 2 in. to 2 ft. depending on habitat and elevation. At higher elevations, it may be only 2 in. tall with small flowers.

FOUND: Moist to wet soil along streams and around springs and seepage areas, or beaver dams from lowest valleys to near timberline throughout the West.

BLOOMS: May into August.

EDIBLE: Leaves.

USES AND FACTS: The leaves can be eaten raw in salad, but have a slightly bitter flavor. This plant grows in alkaline soil around the hot pools at Old Faithful in Yellowstone Park where it is only 2 in. high.

MULLEIN. *Verbascum thapsus* (Figwort family: Scrophulariaceae)
SIZE: 1 to 8 ft. tall. Flowers 1/2 to 3/4 in. broad.
FOUND: Dry, gravelly, or sandy waste areas, roadsides, railroad
grades and occasionally in open forests. Most often found in disturbed
areas in most of temperate North America from sea level to 8,000 ft.
BLOOMS: Late June to August.
USES AND FACTS: Mullein, a European species, has been used as
a medicinal plant for thousands of years. It produces a mild narcotic.
Its tea has a sedative effect, and dried leaves smoked by asthma
patients are supposed to provide relief (Spencer, 1964). Dioscorides
(early Greek physician) said a decoction of roots was given for cramps,
convulsions, coughs and toothache. Also three ounces of distilled
water of the flowers drunk morning and night was a remedy for gout.
Indians dried leaves and smoked them for lung trouble.
Mullein also has astringent properties. The flower oil was used for
earache and coughs (Sweet, 1962). The leaves are supposed to
contain a chemical used in lotions to soften the skin, and in medicines
to sooth inflamed tissues. The plant is a biannual. That is, the first
year it produces a rossette (which can be potted and kept in the house
as a winter greenery), and the next year it blooms, produces seeds,
then dies. The seed stalk keeps well as a winter bouquet.

SWEET CICELY. *Osmorhiza occidentalis* (Carrot or parsley family: Umbelliferae)
SIZE: 1 to 3 ft. high. Flowers 2/5 to 4/5 in. long and in umbels (clusters shaped like umbrella).
FOUND: On wooded slopes and in valleys and meadows from 2,500 to 8,500 ft. from British Columbia to California throughout the West.
BLOOMS: May through July.
EDIBLE: Roots, leaves and seeds. Anise flavor.
USES AND FACTS: Harrington (1967) says the roots of this species are very strongly flavored of anise, and should be dried and powdered then used as flavoring in such things as cookies. Use 2 teaspoons of powdered root and 4-1/2 tablesoons of boiling water. Drain through a cloth or fine strainer to produce a liquid for use in recipes calling for anise. Leaves and seeds can also be used for seasoning, and seeds can be eaten raw.

57

DESERT PARSLEY. *Lomatium* sp. (Parsley or carrot family: Umbelliferae)

SIZE: Species of desert parsley range in size from a few inches high up to 5 ft. tall. Flowers small and in umbels (heads shaped like an umbrella).

FOUND: Dry rocky soils from valleys up to 7,000 ft. in mountains throughout the West.

BLOOMS: April to July.

EDIBLE: All parts.

USES AND FACTS: All species of *Lomatium* have roots that are edible raw (taste like celery), or they may be peeled, dried and ground into flour. Stems may be eaten in spring, but become tough and fibrous later in summer. Tea can be made from the leaves, stems and flowers. The seeds are nutritious raw, roasted or dried and ground into flour (Kirk, 1970).

STONECROP. *Sedum* sp. (Orpine family: Crassulaceae)
SIZE: 4 to 8 in. tall. Flowers 1/4 to 1/2 in. across and in clusters.
FOUND: On rocks or rocky dry soil from lowest valleys to about 9,000 ft. in most Western states.
BLOOMS: Late June to August.
EDIBLE: Stems and leaves.
USES AND FACTS: There are about 300 species of *Sedum* of which 12 are found in the Rocky Mountain area. The fleshy leaves with a waxy cover allow this genus to live in very dry conditions as the waxy cover helps the plant to retain water. Stonecrop can be used for salad or as a potherb. The young stems and leaves have the best flavor. Some species are too acid to be palatable, but none are poisonous. Flowers vary in color from white to yellow or red in different species.

FIDDLENECK. *Amsinckia* sp. (Borage family: Boraginaceae)
SIZE: 11 in. to 2-1/2 ft. in height. Flowers small, growing on a scorpioid-shaped spike.
FOUND: Dry soils of plains, hills, fields and waste areas throughout the West.
BLOOMS: May to July.
USES AND FACTS: The hairs which are abundant on the leaves can pierce the skin and cause irritation. The seeds are reputed to be poisonous to cattle. *Amsinckia* usually grows in disturbed areas.

VIOLET. *Viola purpurea* (Violet family: Violaceae)
SIZE: 4 to 8 in. high. Flowers 1/4 to 1/2 in. long.
FOUND: Dry slopes from 3,000 to 8,000 ft. from Oregon south to California and in other Western states.
BLOOMS: April to June.
EDIBLE: Leaves and flowers.
USES AND FACTS: All species of *Viola* are edible. Young leaves and flower buds are used for salads. Leaves and flowers can also be boiled as potherb. Violets called "wild okra" are used to thicken soup in the southern U.S. The flowers can be candied like rose petals, and have been used to give a flavor to white vinegar. The leaves can be used for tea (Harrington, 1968).

BUCKWHEAT. *Eriogonum sp.* (Buckwheat family: Polygonaceae)
SIZE: Range from mat plants to several feet in height. Flowers small, in heads.
FOUND: Species of buckwheat are found from the deserts at low elevations to the high alpine peaks throughout the Western states.
BLOOMS: May through August, depending on elevation.
USES AND FACTS: There are about 150 species of buckwheat in North America. Because buckwheat has a long blooming season, it is highly utilized by bees. Nectar from buckwheat produces a fine quality of honey. The seeds remain on the plants for a long time, and are excellent fall and winter bird feed. Indians used a decoction from the leaves for headache and stomach pains. Tea from the flowers was used as an eyewash, for high blood pressure and for bronchial ailments. The stems and leaves were boiled for a tea to treat bladder trouble (Sweet, 1962). Buckwheats make excellent dried flowers for winter bouquets. (See white.)

GOLDCUP; YELLOW FRITILLARY. *Fritillaria pudica* (Lily family: Liliaceae)

SIZE: 3 to 12 in. high. Flower about 3/4 in. long, bell-shaped.

FOUND: Grasslands, dry open woods, slopes fields and meadows below 5,000 ft. from British Columbia to California throughout the west.

BLOOMS: April to June.

EDIBLE: Bulb and numerous "rice-grain" bulblets and green seedpods.

USES AND FACTS: Bulbs may be eaten raw or boiled, or can be dried (Kirk, 1970). The green seedpods are delicious cooked or raw (Craighead et. al, 1963). Black, Alaskan brown and grizzly bears, pocket gophers and ground squirrels eat the bulbs. Deer and elk eat the leaves and seedpods. *Pudica* is Latin for ashamed or bashful, suggested by the shyly hanging flower head.

63

PRIMROSE.   *Oenothera heterantha* (Evening primrose family: Onagraceae)
SIZE: 1 to 6 in. high; lies close to ground.   Some species 1 to 2 ft. tall.
Flowers 1/2 to 2 in. broad.
FOUND: Moist grassy places from 6,000 to 9,000 ft. throughout most Western states.   Some species are found on dry soil, steep areas, sandy places, and ridges.
BLOOMS: May to August, depending on species.
EDIBLE: Some species: seeds and roots.
USES AND FACTS: The flowers of all primroses resemble those shown here.   Evening primrose flowers open in the evening and remain open during the day.   The flowers of some species are white.   Indians ate the seeds of several species, and an Eastern species *(O. biennis)* has been cultivated for its tasty and nutritious edible roots.

CINQUEFOIL. *Potentilla* sp. (Rose family: Rosaceae)
SIZE: *P. fruiticosa*, a shrub from 1 to 5 ft. tall. Flowers 3/4 in. broad.
Potentilla sp., 1 to 2 ft. tall. Flowers about 1/2 in. broad.
FOUND: Damp to wet soil from plains to 9,000 ft. throughout the
West. Species of cinquefoil are found in most plant communities in
many kinds of soil.
BLOOMS: Late June to early August, depending on the species and
elevation.
EDIBLE: Leaves.
USES AND FACTS: There are more than 300 species of *Potentilla*,
mostly in the Northern Hemisphere. Both domestic livestock and wild
animals browse *P. fruiticosa* which retains its leaves during winter.
Leaves can be used as a tea. (See white flowers.)

BUTTERCUP. *Ranunculus* sp. (Buttercup family: Ranunculaceae)
SIZE: Species range from 4 in. (alpine buttercup) to 2 ft. in height. Flowers 1/2 to 1 in. broad.
FOUND: Species are found from the desert (sagebrush buttercup) to timberline, mostly in moist soil, even in sagebrush flats, throughout the West.
BLOOMS: April through August.
DANGER: Poisonous. **EDIBLE:** Cooking is supposed to remove toxins.
USES AND FACTS: All members of the buttercup family are more or less poisonous when eaten raw. Delphinium and larkspur belong to this group, as does marsh marigold. The alkaloids delphinine, delphineidine, ajacine and others are found mostly in the seeds and young plants. They cause stomach upset, nervous symptoms, depression, and may be fatal if eaten in large quantities. Danger decreases as the plants age (Hardin and Arena, 1974). Thorough cooking (Kirk, 1970) is supposed to remove the toxins. Western Indians parched seeds and ground them into meal for bread. Roots were boiled and eaten, as were leaves of some species. Early Western settlers pickled the young flowers. Some Indians made yellow dye by crushing and washing the flowers. There are about 35 species of buttercup in the Western states.

66

WAYSIDE GROMWELL. *Lithospermum ruderale* (Borage family: Boraginaceae)
SIZE: 1/2 to 1-1/2 ft. high. Flowers 1/3 in. wide.
FOUND: Dry slopes and plains from 4,500 to 6,000 ft. in sagebrush scrub and juniper woodland throughout the West.
BLOOMS: May to June.
USES AND FACTS: The roots of some members of this genus were cooked and eaten by Western Indians. The plants were used as both medicine and food. *L. ruderale* is supposed to have been used as a contraceptive by Indians and early pioneers.

TIGER LILY. *Lilium columbianum* (Lily family: Liliaceae)
SIZE: 2 to 5 ft. high. Flowers 1 to 2 in. long.
FOUND: Prairies, thickets, conifer forests from sea level up to 6,000 ft. and higher in mountains from British Columbia to Northern California.
BLOOMS: June and July.
EDIBLE: Corm or "bulbs."
USES AND FACTS: The corms taste like bitter roasted chestnuts (Turner, 1975), but can be eaten raw or cooked, usually steamed. However, because this is a relatively rare and beautiful plant, we recommend leaving them alone unless in an emergency. Many legends have been inspired by lilies. One was that of a Korean hermit who removed an arrow from a wounded tiger and became friends with it (Kirk, 1970). When the tiger died, his body was transformed into a tiger lily so that he could remain close to the hermit. Years later, when the hermit drowned and was washed away, the lily spread throughout the land seeking his friend.

MOUNTAIN-SORREL. *Oxyria digyna* (Buckwheat family: Polygonaceae)
SIZE: 6 to 12 in. tall. Flowers small, greenish; fruits reddish-brown.
FOUND: In shady, wet or moist places generally on mountain slopes, ledges and rock crevices from 6,000 to 11,000 ft. throughout the West.
BLOOMS: Late June through August.
EDIBLE: Leaves and stems.
USES AND FACTS: The leaves and stems have a pleasant sour taste when eaten raw in salads, and are also good when boiled. Some Indian tribes, especially in Canada and Alaska, ferment the leaves slightly as a sort of sauerkraut. Others store it for winter use (Kirk, 1970). Mountain sorrel is high in vitamin C, and was used in early times to prevent and cure scurvy. Elk eat mountain sorrel.

RUMEX SP.

RUMEX SP.

DOCK or **SORREL.** *Rumex* sp. (Buckwheat family: Polygonaceae)
SIZE: Varies with the species, ranging from 1 to 3 ft. tall. Flowers small and greenish; fruits, reddish-brown.
FOUND: Species are found throughout temperate North America in moist soil along roads, irrigation ditches, pastures, cultivated fields and waste lands from valleys to about 6,000 ft. in the mountains.
BLOOMS: June to July.
EDIBLE: Leaves and stems.
USES AND FACTS: Kirk (1970) says that all species of *Rumex* have edible leaves and stems, but some have less acid than others. Those that are particularly tart or bitter should be boiled two or three times in fresh water which will remove most of the acid, but won't destroy the flavor. Just a small amount of water should be used for the last boiling so that the leaves and stalks won't be too watery and unpleasant when eaten. Some species need be boiled only once, depending on individual taste. *R. crispus*, curlydock, has a root that was formerly used as a drug for laxative and tonic, and was sold under the name "yellowdock." Navajo Indians extracted a dye from the roots of *R. hymenosepalus*.

71

FALSE DANDELION. *Agoseris aurantiaca* (Sunflower family: Compositae)
SIZE: 4 to 12 in. tall. Flowers about 1 in. broad.
FOUND: Moderately dry to moist, or even wet soils of meadows and open areas in mountains at most elevations throughout the Rocky Mountains.
BLOOMS: May into August, depending on elevation.
USES AND FACTS: About a dozen species of *Agoseris* occur in the Rocky Mountains. *Agoseris* have a milky juice or sap which, when solidified, was reportedly chewed by Western Indians. It is moderately grazed by livestock, and domestic sheep are very fond of it.

ORANGE GLOBE MALLOW. *Sphaeralcea munroana* (Mallow family: Malvaceae).

SIZE: 4 to 18 in. tall. Flowers about 1 in. broad.

FOUND: In many soil types, but "prefers" moderately sandy or rocky sites of plains, valleys, foothills and mountains to about 6,000 ft. throughout Western states.

BLOOMS: May to August, depending on elevation.

EDIBLE: Leaves and seeds.

USES AND FACTS: *Spaera* is Greek, meaning globelike and refers to the round fruit with pie-shaped segments; *alcea* means mallow. This plant closely resembles the domestic geranium. Cotton, hollyhocks, mallows, okra and many other garden flowers belong to the mallow family. The leaves and stems of all species may be boiled as a green, but one species, *Malva neglecta*, is the most flavorful. However, large amounts should not be eaten at any one time because they will cause a digestive disorder (Kirk, 1970). *M. neglecta* is a common weed found in gardens and waste areas. Its seeds, which look like little round cheeses, are edible and nutritious. This species closely resembles orange globe mallow, but has smaller flowers that are pinkish or whiteish.

73

SITKA COLUMBINE. *Aquilegia formosa* (Buttercup family: Ranunculaceae)

SIZE: 20 to 40 in. high. Flowers about 1½ in. long.

FOUND: In moist areas from 4,000 to 9,000 ft. in aspen woodlands and other plant communities throughout the Rockies.

BLOOMS: June to August.

USES AND FACTS: Nine different species of columbine occur in the Rocky Mountain area. One, *A. coerulea*, has a blue and white flower, and is the state flower of Colorado. Pure white and pure yellow species are also found, but all flowers resemble those of the Sitka columbine. Columbine are important forage plants, but on overstocked range, domestic sheep graze them heavily, and all species are becoming rare in areas where once they were abundant (Craighead, *et al.*, 1963). Wild columbine look like and are related to the domestic columbine.

SCARLET GILIA. *Gilia aggregata* (Phlox family: Polemoniaceae)
SIZE: 1 to 3 ft. tall. Flowers 3/4 to 1-1/2 in. long.
FOUND: Dry soil of lowest valleys, hillsides and mountain ridges up to timberline from British Columbia to California throughout the Western states.
BLOOMS: May to July.
DANGER: May be poisonous.
USES AND FACTS: The blossoms of this plant are usually red, but they vary from white to pink and orange. The upper leaves have a distinctive skunk odor when crushed, so the plant is somtimes called "Polecat Plant." *Gilia* is reported to contain saponin, a chemical allied to soap, and is poisonous (Craighead, *et al.*, 1963). However, sheep eat this plant without ill effects. This beautiful flower can frequently be seen growing in large patches in mountain meadows.

PAINTBRUSH. *Castilleja* sp. (Figwort family: Scrophulariaceae)
SIZE: 4 to 16 in. tall. Flowers greenish, up to 1 in. long, but often concealed by brightly colored bracts that are mistaken for flowers.
FOUND: Dry, medium-dry to moist soils from the deserts to timberline in most Western states.
BLOOMS: May to September, depending on species and elevation.
EDIBLE: Flowers. *CAUTION:* May be toxic if eaten in large quantities.
USES AND FACTS: Paintbrush species ranging in color from white, yellow, orange, pinkish to reddish can be found often growing in large patches in mountain meadows or other habitats. They are extremely colorful because of the colored bracts that surround the greenish flowers. They often look like they have been dipped in paint, and in late afternoon the red and orange species give the landscape an appearance of being afire. Most, if not all, species have edible flowers that can be eaten raw (Kirk, 1970). However, when the plant is growing on soil that contains selenium, it can absorb this element which may be toxic. Therefore, care should be exercised, and only small quantities should be eaten. Paintbrush is semi-parasitic and grows with its roots close to those of another plant, obtaining part of its nutrition from the neighboring plant. There are 200 species of paintbrush in temperate North America. *C. linariaefolia* is the state flower of Wyoming.

76

CLOVER. *Trifolium* sp. (Pea family: Leguminosae)
SIZE: 4 to 12 in. high. Flowers in heads.
FOUND: Wet meadows, along streams and roadsides and in mountains up to about 9,000 ft. throughout the West.
BLOOMS: Late May through July.
EDIBLE: Leaves, seeds and flowers.
USES AND FACTS: All species of clover can be eaten raw (Kirk, 1970) but should be eaten sparingly as they are hard to digest, and can cause bloat. They are high in protein and nutritious, and can be eaten in quantity if cooked or soaked for several hours in strong salt water. The seeds and dried flowers can also be used for food. A tea may be made by steeping the dried flowers for a few minutes in hot water. Clovers are valuable forage crops, and are important summer food for ruffed and sage grouse, Canada geese, deer, elk, black and grizzly bears (Craighead, *et al.*, 1963). Red clover is Vermont's state flower.

WILD PEA. *Lathyras* sp. (Pea family: Leguminosae)
SIZE: Species of pea range from erect plants of several feet in height to vines which twine over other plants. The flowers are about 1 in. long and resemble those of the cultivated sweetpea.
FOUND: Species are found in all types of soils from about 4,000 ft. to above 7,000 ft. throughout the West.
BLOOMS: April to July.
DANGER: Some species edible, others are toxic.
USES AND FACTS: Kirk (1970) lists *L. japonicus* as a species whose young pods and seeds are edible. However, he says that other species are toxic. "Evidently these plants have a history of poisoning humans dating back to Grecian times." He says that there have been epidemics of "lathyrism" reported, usually during drought or poverty conditions which have encouraged people to eat large quantities of wild peas. Eaten daily in normal amounts, many species of this plant are nutritious food. However, if eaten as an exclusive diet for ten or more days, some species have caused a partial or total paralysis, which is permanent, accompanied by a variety of secondary disorders.
"Even the common garden pea [*Pisum*], closely related to *Lathyrus*, can cause nervous disorders if eaten in great quantities over long periods of time."

79

TWIN FLOWER. *Linnaea borealis* (Honeysuckle family: Caprifoliaceae)
SIZE: Small, mat-forming evergreen, herb-like shrubs that often carpet forest floor. Flower stem 2 to 6 in. high with paired, bell-shaped flowers about ½ in. long.
FOUND: In woods, forests from lowest elevations to about 9,000 ft. throughout the West and east to Minnesota, south to West Virginia.
BLOOMS: June and July.
EDIBLE: Leaves for tea.
USES AND FACTS: The Snohomish Indians boiled the leaves for a tea to treat colds (Gunther, 1981). This plant was named for Carolus Linnaeus of Sweden, the man mostly responsible for the binomial nomenclature system (generic and specific names) of plants and animals. The two Latin words constitute the "scientific" name of an organism and that name is consistent throughout the world, unlike common names which are often quite localized. Twin flower has a small, very sticky seed that clings to small animals and the clothing of people and is disseminated in that manner. Masses of blooming twin flower fill the air of the forest with a delicate and lovely scent.

DUSTY MAIDEN. *Chaenactis douglasii* (Sunflower family: Compositae)

SIZE: 4 to 18 in. tall. Flowers small, in heads.

FOUND: Dry to medium-moist soil along roadsides, waste places and hillsides, especially where soil has been distrubed, from lowest valleys to timberline from British Columbia to California throughout the West.

BLOOMS: June to early August.

USES AND FACTS: There are 11 species of *Chaenactis* found in the Rocky Mountains area. The species look much alike and are difficult to distinguish from one another. *Chaenactis* is sometimes confused with yarrow, but the dusty maiden does not have ray flowers as does the yarrow.

ELK THISTLE. *Cirsium foliosum* (Sunflower family: Compositae)
SIZE: Stems from 2 or 3 in. up to 4 ft. high. Flower head ½ to 2 in. broad.
FOUND: Moist to wet soil of open meadows and valleys up to about 8,000 ft. from British Columbia to California and New Mexico.
BLOOMS: June to early August.
EDIBLE: Roots and stems.
USES AND FACTS: Although somewhat flat in taste, the roots and stems of all thistles are edible and nutritious. Since thistles are easy to recognize, they make good emergency food. The roots can be eaten raw, boiled or roasted. Peeled stems can be cooked or eaten raw as greens. Thistledown from the seeds makes a good tender for fire starting (Kirk, 1970; Craighead, *et al.*, 1963).

STICKY GERANIUM. *Geranium viscosissimum* (Geranium family: Geraniaceae)
SIZE: 1 to 2 ft. tall. Flowers about 1 in. broad.
FOUND: Medium-dry to moist or even wet soil of open woods, roadsides, stream banks and meadows from valleys to 9,000 ft. from British Columbia to California throughout the West.
BLOOMS: May to July or August, depending on elevation.
USES AND FACTS: This geranium belongs to the same family as the domestic geranium, but to a different genus. There are about nine species of wild geranium in the Rocky Mountain area, some with white flowers. Sticky geranium is a valuable forage plant, and is abundant over much of the Western range land. It is a major food for elk and deer in the spring and summer, and is also eaten by black bears and probaby by grizzlies. Moose will choose the flowers and upper leaves in preference to other vegetation (Craighead, *et al.*, 1963).

MINT: *Mentha piperita* (Mint family: Labiatae)
SIZE: 1 to 3 ft. tall. Flowers small, in spikes.
FOUND: Wet soil of streambanks, around springs, bogs, along roads and in wet woods from lowest elevations to about 8,000 ft. throughout temperate Northern Hemisphere.
BLOOMS: July to September.
EDIBLE: Leaves.
USES AND FACTS: This species of mint is one that came from the Old World and has escaped cultivation. It has become widely naturalized in the United States. Peppermint used in medicine and for flavoring comes from this plant. Tea can be made by steeping the fresh or dried leaves in hot water. Menthol is derived from a cultivated variety of *M. arvensis*, the native species of mint. All mints can be used for making jelly or mint juleps.

MOUNTAIN HOLLYHOCK. *Iliamna rivularis* (Mallow family: Malvaceae)

SIZE: 3 to 6 ft. tall. Flowers 1 to 2 in. broad.

FOUND: Rich, moist soil along streams, in canyons, on roadsides and in open areas from foothills to almost 9,000 ft. from British Columbia south to Nevada and Colorado. Found in Idaho and other northwestern states.

BLOOMS: June to early August.

DANGER: Tiny hairs on the ripe fruit are irritating to the skin.

USES AND FACTS: Mountain hollyhock flowers are somewhat smaller than, but similar to those of domestic hollyhocks. This plant would make an attractive garden plant, and can be started from the seeds. There are three species of hollyhock in the Rocky Mountain area. The Indians chewed the stems of some hollyhocks as gum.

PRAIRIE SMOKE. *Geum triflorum* (Rose family: Rosaceae)
SIZE: 6 to 14 in. high. Flowers about 1/2 in. long.
FOUND: Medium-dry soil of hillsides and ridges to over 8,000 ft. from British Columbia throughout the West to California.
BLOOMS: May to July.
EDIBLE: Roots.
USES AND FACTS: Indians boiled the roots to make a tea that tastes somewhat like weak sassafras tea. *Triflorum* means three-flowered. The seeds of this plant have long feathery tails that act like a sail and help scatter the seeds. Since the seeds are often more conspicuous than the flowers, we have included a picture of them. The fern-like leaves are one of the first green things to appear as the snow recedes (Craighead, *et al.*, 1963).

SCABLAND PENSTEMON. *Penstemon deustus* (Figwort family: Scrophulariaceae)
SIZE: 8 to 24 in. tall. Flowers 1/2 in. long.
FOUND: Dry, rocky places below 8,200 ft. in desert and mountains. Found at Craters of the Moon National Monument.
BLOOMS: May to July.
USES AND FACTS: One of the many species of penstemon found throughout the West. This particular plant is showy and attractive, but has a particularly displeasing odor on warm days. (See blue, lavender flowers.)

WILD ONION. *Allium* sp. (Lily family: Liliaceae)
SIZE: 6 to 18 in. tall. Flowers small, in clusters.
FOUND: Species of onion are found in dry to moist soils from the deserts to the high mountains throughout the West.
BLOOMS: June to August, depending on elevation.
EDIBLE: Entire plant. ***DANGER:*** Toxic if eaten in huge quantities.
USES AND FACTS: All of the onion species are edible, but their bulbs vary in degree of onion flavor and odor. They can be eaten raw, boiled, steamed, creamed, in soup or stew, and are especially good when used as a seasoning. They store well for winter use. It should be remembered that eating huge amounts of onions, including domestic onions, can cause poisoning. It is thought that onions contain an alkaloid, but it is safe to eat normal amounts of wild onions in the same manner as you would eat cultivated ones (Kirk, 1970). Wild onions vary in color from white to pink to red.

LUPINE. *Lupinus sp.* (Pea family: Leguminosae)
SIZE: From 2/3 to 3-1/2 ft. tall. Flowers vary in size and color.
FOUND: Species are found from deserts to above timberline in moist and dry soils throughout the West.
BLOOMS: Late June to early August, depending on species and elevation.
DANGER: Poisonous.
USES AND FACTS: Lupines are an important group of plants that are especially dangerous to sheep. The fruits and seeds are considered to be the most dangerous parts, but the foliage is also dangerous until after the seeds are ripe. A lethal dose must be eaten at one time to cause death. Sweet (1962) refers to Europeans and Indians using lupine. She says, "Virgil called it 'sad lupine' as seeds were used by the poor who boiled them to remove the bitter taste." A species of lupine (probably *L. albus*) closely related to our native species is raised in Europe for the edible seeds which are canned and are sometimes exported to this country. However, the alkaloids in our lupine make them very dangerous, and we strongly recommend avoidance of the seeds, especially in an emergency. Cooking may remove the poison, but there is no positive evidence of this.

STRAWBERRY BLITE; BLITE GOOSEFOOT. *Chenopodium capitatum* (Goosefoot family: Chenopodiaceae)
SIZE: 8 in. to 2½ ft. tall. Flowers small, greenish and inconspicuous, developing into red and fleshy fruit resembling a berry.
FOUND: In moist, rocky soil in the mountains under 10,000 ft. throughout most of the United States.
BLOOMS: June to August.
EDIBLE: Red, fleshy fruit clusters; leaves and tender shoots.
USES AND FACTS: *Chenopodium* means "goosefoot," descriptive of the leaves. Beets and spinach belong to the same family as strawberry blite. Leaves and tender shoots of the plant can be used as a potherb, and the fruit clusters can be eaten raw or cooked, although it is rather bland and seedy. *C. capitatum* is an introduced species from Eurasia. Its tendency to grow in disturbed areas indicates that it is not a native plant. *C. album*, a related species commonly called pigweed or lambs quarters, is also edible, as are other species of *Chenopodium*. Pigweed is often found growing in gardens and disturbed areas. The seeds, young leaves and tender shoots are all considered edible.

SHOOTING STAR. *Dodecatheon pauciflorum* (Primrose family: Primulaceae)

SIZE: 6 to 16 in. tall. Flowers 1/2 to 1 in. long.

FOUND: Moist to wet soil in open places of plains, meadows, hills and mountain sides from lowest valleys to nearly 12,000 ft. from Canada throughout the West to California.

BLOOMS: Late April through July, depending on elevation.

EDIBLE: Some species: roots and leaves.

USES AND FACTS: The roots and leaves of *D. hendersonii* can be eaten roasted or boiled. Since none of the species are listed anywhere as poisonous, it is likely that all are edible (Kirk, 1970). However, extreme caution should be used with species other than *D. hendersonii*, because no one seems to know for sure about the edibility of the other species. Elk and deer eat shooting stars in the early spring when other green forage is still scarce.

LEWIS' MONKEYFLOWER.   *Mimulus lewisii* (Figwort family: Scrophulariaceae)

SIZE: 1 to 1-1/2 ft. high.   Flowers 1 in. broad.

FOUND: Occurs only in wet places from 5,000 to 10,000 ft. from British Columbia to California throughout the West.

BLOOMS: Late June through August.

EDIBLE: Stems and leaves.

USES AND FACTS: Sweet (1962) says that Indians used both young stems and leaves for salad greens. The root of yellow *Mimulus* was used as an astringent. Raw leaves and stems were crushed and applied to rope burns and wounds as a poultice. The common name "monkeyflower" refers to the "grinning face" of the variously colored flowers that resemble the masks worn by comic stage actors. The plant has little forage value, but is used occasionally by mountain sheep, elk and deer, (Craighead, *et al.*, 1963).

DWARF MONKEYFLOWER. *Mimulus nanus* (Figwort family: Scrophulariaceae)
SIZE: Rarely exceeding 2 in. in height. Flowers about 1/2 in. wide.
FOUND: Bare areas and loose soils from desert to mountains throughout the West.
BLOOMS: June to August.
USES AND FACTS: This delightful little flower can be seen growing individually or in patches. It is one of the 20 species of monkeyflowers that occur in the Rocky Mountain area. Since the leaves of the yellow monkeyflower are edible. it is probable that other species can be eaten too. However, spare this tiny, hearty little flower that grows under adverse conditions and is a delight to behold.

ELEPHANTHEAD.   *Pedicularis groenlandica* (Figwort family: Scrophulariaceae)
SIZE: 8 to 24 in. high. Flowers small, in spikes.
FOUND: Wet soil of bogs, meadows and along streams and lakeshores, often growing in shallow water, and usually in open places from about 5,500 ft. to above timberline from Alaska to California throughout the West.
BLOOMS: June to August.
USES AND FACTS: Often seen growing in boggy meadows in patches, this is one of the most showy wildflowers to be found. The flowers resemble a pink elephant head, and cannot be confused with any other plant. The name *pedicularis* is Latin for louse, and comes from an old superstition that eating these plants increased the lice on cattle. Nine species of *Pedicularis* occur in the Rocky Mountain area and almost 500 species in the temperate regions of the world. The yellow roots of an arctic species (*P. lanata*) taste somewhat like carrots, and can be eaten raw or cooked. Elk graze elephanthead in early summer (Craighead, *et al.*, 1963).

FIREWEED. *Epilobium angustifolium* (Evening primrose family: Onagraceae)

SIZE: 1 to 7 ft. tall. Flowers almost 1 in. across.

FOUND: Moist, rich soil in open woods, prairies, hills, especially along streams, in damp places and on disturbed ground; areas where forest fires have occurred from lowest valleys to treeline throughout the West.

BLOOMS: June through August.

EDIBLE: Young leaves and shoots.

USES AND FACTS: Nearly 25 species of *Epilobium* are found in the Rocky Mountains. This plant is called fireweed because after forest fires, it invades the burned areas, covering the scar and providing valuable forage for both livestock and wild game such as deer and elk. It is also a favorite forage of grizzly bears. The young shoots and leaves may be boiled as a potherb or eaten raw in salad. The leaves, green or dry, make good tea. The pith of the stems can be used in soup.

95

WILD PEONY. *Paeonia brownii* (Peony family: Paeoniaceae)
SIZE: 8 to 16 in. high. Flower 1 to 1½ in. across.
FOUND: Dry slopes from 3,000 to 7,500 ft. from British Columbia to California throughout the Western states.
BLOOMS: April through May.
USES AND FACTS: Little information is available about the wild peony. However, Paeon is Greek, meaning physician of the gods, which might suggest that the plant had some ancient medicinal use. There are 30 species of peony in the Northern Hemisphere, largely Asian. Many are of horticultural value, and numerous varieties of domestic peonies can produce a colorful, showy garden in the springtime. The peony is the state flower of Indiana.

SKUNKWEED. *Polemonium* sp. (Phlox family: Polemoniaceae)
SIZE: 10 to 20 in. tall. Flowers 3/4 to 1 in. long.
FOUND: Wet places from 3,000 to 10,000 ft. throughout the West.
BLOOMS: June to mid-August.
USES AND FACTS: There are about 11 species of *Polemonium* in the Rocky Mountain region. This plant has a powerful skunk odor that will cling to your shoes if you should step on it. The plant is deceptive because it is so attractive, and, should you see it, you will no doubt bend over to smell it. Are you in for a shock! The odor is probably the plant's protection against grazing animals.

FLEABANE. *Erigeron* sp. (Sunflower family: Compositae)
SIZE: Species range in size from 4 to 24 in. tall. Ray flowers (those looking like petals) can be white, pink, blue or lavender. Disc flowers (in center) are usually yellow.
FOUND: In all soil types from low deserts to above timberline throughout the West.
BLOOMS: May through August, depending on elevation and species.
USES AND FACTS: See white fleabane.

ASTER SP.

ASTER: *Aster* sp. (Sunflower family: Compositae)
SIZE: 2 to 5 in. high. Flower heads about 1 in. across.
FOUND: Dry, rocky hillsides from 5,000 to 10,000 ft. throughout the Western states.
BLOOMS: May to July.
USES AND FACTS: There are about 35 species of *Aster* found in the Rocky Mountain area. Some species of aster bloom until late fall and are around after many other species of flowers have gone to seed. Asters provide fall and winter feed for elk.

SUGARBOWL.   *Clematis   hirsutissima*   (Buttercup   family:
Ranunculaceae)
SIZE: 1 to 2 ft. tall. Flowers sugarbowl-shaped, about 1 in. long and
3/4 in. broad.
FOUND: In moist open areas of plains, hills and woods to about
8,000 ft. in the mountains in Montana, British Columbia, Idaho,
Oregon and Colorado, and other Northwestern states.
BLOOMS: Late April through June.
USES AND FACTS: *Hirsutissima* means very hairy. Indians are
reported to have used this plant medicinally. The feathery seeds of
Clematis can be used as a floral decoration. The seed heads, which
are pictured here, are as showy as the flower.

CALYPSO ORCHID; FAIRYSLIPPER. *Calypso bulbosa* (Orchid family: Orchidaceae)
SIZE: Flower stem 2 to 7 in. high. Single basal leaf. Flowers showy, about 1 to 1½ in. long.
FOUND: Evergreen forests throughout the West from Alaska to California in deep shady moist areas, often near decayed stumps and logs.
BLOOMS: Late April to June, soon after snow melts and depending on rainfall. Blooming season shorter when dry.
EDIBLE: Bulb.
USES AND FACTS: The small bulb is quite palatable when eaten raw, roasted or boiled. DON'T eat it unless in an emergency because this beautiful flower is quite rare and should be spared. Although there are seldom a lot of orchids in one locality, the orchid family is one of the largest in the world with from 8,000 to 10,000 species. Orchids are adapted for cross pollination by insects and even when seeds are produced, the orchids will frequently not germinate unless stimulated by the presence of certain fungi (Craighead, *et al.*, 1963). It is important, therefore, that orchids not be picked (they wilt fast), or their bulbs destroyed, because each flower or bulb removed will reduce the chance of further production of this exotic species.

102

LEOPARD LILY. *Fritillaria atropurpurea* (Lily family: Liliaceae)
SIZE: 8 to 30 in. tall. Flower about 1 in. wide.
FOUND: In leaf mold under trees; in rich damp soil of valleys, woods and in mountains from 6,000 to 10,500 ft. in montane forests throughout the Western states.
BLOOMS: Late April to July.
EDIBLE: Corms.
USES AND FACTS: *Atropurpurea* means dark purple. The flower is dark purple with yellowish-green spots. It hangs down on the plant so is frequently missed because the color blends into the surroundings. Look carefully for it as it is worth seeing. The starchy corms (bulb-like roots) are edible and are still eaten by Western Indians and Eskimos. Use it sparingly, if at all, and only in an emergency as these beautiful flowers are not too common.

103

DWARF FORGET-ME-NOT. *Eritrichium elongatum* (Borage family: Boraginaceae)

SIZE: Plant forms a cushion rarely more than 4 in. high. Flowers less than ¼ in. broad.

FOUND: Only at high altitudes up to 12,000 ft. on exposed ridges and mountain crests from Montana, Idaho and Oregon south to New Mexico.

BLOOMS: July to August.

USES AND FACTS: There are about 30 species of *Eritrichium* in the Northern Hemisphere, and similar species are found in the Alps. The white flower shown with the forget-me-not is phlox (*Phlox caespitosa)* which is found in the same area, also in a matt-like growth. Sometimes forget-me-nots will be white, but they can easily be distinguished from phlox which has a larger flower. Also, the leaves and stems of the forget-me-not have white hairs which produce a distinctive grayish appearance, as opposed to the green of the phlox leaves. The Greek root *erion* means wool, and *trichos,* hair. This lovely little flower presents a breath taking sight! The true forget-me-not is Alaska's state flower.

COMMON OR BLUE CAMAS. *Camassia quamash* (Lily family: Liliaceae)

SIZE: 1 to 2 ft. tall. Flowers about 3/4 in. broad, in spikes.

FOUND: In wet meadows and along streambanks from Canada to California throughout the West.

BLOOMS: Middle April to mid-June.

EDIBLE: Bulb.

USES AND FACTS: Camas was an important food source of Indians, trappers and early settlers in the West. The bulbs of the plant are starchy, have a high sugar content, and are nutritious. They can be eaten raw, baked, boiled, roasted or dried. Boiled bulbs have a potato-like flavor, but are slightly slimy or gummy and less mealy than potatoes. Many Indian wars were fought over collecting rights to certain camas meadows (Craighead *et al.*, 1963). Although the bulbs are edible at any season, it is wise to collect them during the blooming season to avoid confusing them with the poisonous bulbs of the death camas (*Zigadenus* sp.) which they closely resemble.

105

LUPINE. *Lupinus* sp. (Pea family: Leguminosae)
SIZE: From 2/3 to 3½ ft. tall. Flowers vary in size and color.
FOUND: Species found from deserts to high mountains in moist and dry soils throughout the West.
BLOOMS: Late June to early August.
DANGER: Poisonous.
USES AND FACTS: Sweet (1962) says that Indians made a tea from the seeds which was used medicinally to help urination. Early in spring, Indians stripped off the leaves and flowers, steamed them, and used them mixed with acorn soup. However, Sweet warns that the seeds are often dangerous because of alkaloids. *L. subcarnosus*, Texas bluebonnet, is the state flower of Texas.

WILD IRIS. *Iris missouriensis* (Iris family: Iridaceae)
SIZE: 1 to 2 ft. high. Flowers 2 to 3 in. long.
FOUND: In open wet meadows from lowest valleys to about 9,000 ft. from British Columbia to California throughout the West.
BLOOMS: May through July.
DANGER: Poisonous.
USES AND FACTS: The rootstocks contain the poison irism which is a violent emetic and cathartic. When the seeds are eaten, they cause painful "burning" of the mouth and throat which persists for several hours. The iris rhizome (rootstock) is suspected of causing dermatitis in some people (Hardin & Arena, 1974). The roots were ground by Indians, mixed with animal bile, then put in a gall bladder and warmed near a fire for several days after which arrow points were dipped in the mixture. It is reported by old Indians that many warriors, only slightly wounded by such arrows, died within 3 to 7 days. Iris are good indicators of water near the surface. The iris (fleur-de-lis) is the emblem of France and the state flower of Tennessee.

BLUE-EYED GRASS. *Sisyrinchium sarmentosum* (Iris family: Iridaceae).

SIZE: 5 to 12 in. tall. Flowers about 3/8 in. long.

FOUND: Wet open areas from lowest valleys to near 8,000 ft. throughout the West.

BLOOMS: May to July, depending on elevation.

USES AND FACTS: One frequently overlooks the blue-eyed grass, but once spotted, it seems to be everywhere. It frequently grows in areas where wild iris is growing, so look closely for it; it is well worth seeing. It is very possible that this plant is poisonous since it is a member of the Iris family and iris are poisonous.

WILD FLAX. *Linum lewisii* (Flax family: Linaceae)
SIZE: 8 to 24 in. tall. Flowers 1/2 to 1 in. across.
FOUND: Dry plains, hills and open ridges from low valleys to about 8,000 ft. throughout the West.
BLOOMS: June to early August.
DANGER: Seeds contain cyanide; causes drowsiness in livestock when eaten.
EDIBLE: Seeds are safe to eat after cooking.
USES AND FACTS: Indians ate the seeds, which have a high oil content, roasted, dried and ground, or cooked with other foods. Linen is made from flax fibers, and many Egyptian mummies were wrapped in fibers made from flax. The stems of numerous wild plants, including flax, were used by Indians for making cordage, varying from ropes to fishing lines. Linseed oil, obtained from flax seed, is used in paints, oilcloth, printer's ink, lineoleum and varnishes. Flax is used medicinally as a laxative, an applicant for burns and scales and a poultice (Craighead, *et al.*, 1963; Kirk, 1970).

PENSTEMON SPECIOSUS

GREAT BASIN PENSTEMON

ROCK PENSTEMON

PENSTEMON. *Penstemon speciosus.* *Penstemon* sp. (Figwort family: Scrophulariaceae)

SIZE: *P. speciosus* up to 3 ft. high. Flowers about 2 in. long. Other species range from 6 in. to 2 ft. in height and flowers vary in size up to 1½ in. long.

FOUND: Species of penstemon are found in fairly moist soils from the deserts to above timberline in the mountains throughout the Western states.

BLOOMS: May through August, depending on species and elevation.

USES AND FACTS: All penstemons can be recognized by their general appearance. The flowers all have a tubular corolla (the petals fused together) with a 2-lobed upper lip and a lobed lower lip. The leaves are always opposite each other on the stem. Species range in color from white to pink, lavender, red, blue and purple. They are beautiful and showy plants. Sweet (1962) says that Spanish New Mexicans boiled the flowering tops and drank the liquid for kidney trouble. Indians made a wash and poultice for running sores, and also steeped the tops for colds. Red Penstemons were boiled, and the solution was used as a wash for burns which was supposed to stop pain and help new skin to grow.

111

MOUNTAIN BLUEBELL. *Mertensia ciliata* (Borage family: Boraginaceae)
SIZE: 1 to 4 ft. high. Flowers about 1 in. long.
FOUND: Species are found from the desert to high mountains in moist soils throughout the Rocky Mountain region.
BLOOMS: Early June to middle August, depending on elevation.
USES AND FACTS: These very showy bluebells can be observed growing in clumps or pure stands in mountain meadows. Elk not only graze bluebells, but also bed down in them and frequently give birth in meadows of bluebells. Deer and bear feed on the entire plant, and domestic sheep are particularly fond of it. Pikas, small rock rabbits, store bluebells for winter use (Craighead, *et al.*, 1963).

MOUNTAIN GENTIAN. *Gentiana calycosa* (Gentian family: Gentianaceae)

SIZE: 4 to 15 in. tall. Flowers 1-1/2 to 1-3/4 in. long.

FOUND: On rocky outcrops, moist slopes, banks of streams and in mountain bogs from 7,000 to 10,000 ft. throughout the West.

BLOOMS: Late July into September.

EDIBLE: Leaves as tea.

USES AND FACTS: Relatives of this species of gentian are found in the Alps, Himalayas and Andes, and are very similar in appearance. Some species grow up to 16,000 ft. elevations in the Himalayas. Gentians were used in Europe and Asia for their medicinal value as they contain a clear bitter fluid that is supposed to have a tonic effect. Early American settlers used some American species in a similar fashion (Craighead *et al.*, 1963). Gentian tea is still used as a tonic in the Alps.

113

MONKSHOOD. *Acontium columbianum* (Buttercup family: Ranunculaceae)

SIZE: 2 to 5 ft. tall. Flowers about 1 in. long.

FOUND: In wet meadows, along streams and springs, from 6,000 to 9,000 ft. from Montana and British Columbia to California throughout the West.

BLOOMS: Late June to early August.

DANGER: Poisonous.

USES AND FACTS: This plant is virulently poisonous to livestock, and has occasionally caused death to humans. The entire plant is poisonous, containing the alkaloids aconine and aconitine. The roots which are the most potent can easily be mistaken for certain edible fleshy roots. The plant is poisonous both before and after flowering. When chewed raw, it causes a tingling sensation in the mouth. The drug aconite, used as a sedative, is derived from a European monkshood (Harrington, 1968; Horn, 1972).

114

LARKSPUR. *Delphinium* sp. (Buttercup family: Ranunculaceae)
SIZE: 6 in. to 3 ft. tall. Flowers 1/2 to 3/4 in. long.
FOUND: Species are found in dry and wet soil from deserts to high mountains throughout the West.
BLOOMS: April through July.
DANGER: Poisonous.
USES AND FACTS: Larkspur is related to domestic delphinium which is also poisonous. They range in color from white to purple. The alkaloids delphinine, delphineidine, ajacine and others are found mostly in the seeds and young plants. These alkaloids cause stomach upset, nervous symptoms, depression and may be fatal if eaten in large quantities. Danger decreases as the plant ages (Hardin, 1974). Larkspur is responsible for the greatest cattle loss on national forest range land, and is deadly to cattle and horses, but not usually to sheep. Elk appear to avoid larkspur in early spring, but feed heavily on it in late summer and fall. The plants seem to lose their toxicity after they have bloomed (Craighead, *et al.*, 1963). Most members of the buttercup family are poisonous and should be avoided as food.

PASQUE FLOWER. *Anemone occidentalis* (Buttercup family: Ranunculaceae)
SIZE: Flowers 1 to 1½ in. across, usually singly at end of 2 to 16 in. stem. Leaves basal.
FOUND: Dry rocky slopes in montane conferous forest from mid-mountain to sub-alpine meadows, fields and woods of Western states.
BLOOMS: Late June, July and August.
DANGER: Poisonous; entire plant.
USES AND FACTS: The color of the flower varies from white to pale blue to purplish. This is the state flower of South Dakota. The seeds of this plant have long persistent styles that look long and feathery and resemble those of sugarbowl. All anemones are poisonous, containing protoanemonin. Ingestion of the plant is usually limited because of irritation to skin and mucous membrane, often producing blisters. Chewing causes acute inflamation and occasional ulceration of the mouth and throat. If swallowed, severe gastroenteritis, bloody vomite and diarrhea occur (Lampe & McCann, 1985). Domestic sheep have died from overeating this plant, although it is usually avoided when other forage is available because it tastes acrid and is hairy (Craighead, *et al.*, 1963).

116

SILKY PHACELIA. *Phacelia serecia* (Waterleaf family: Hydrophyllaceae)

SIZE: 5 to 18 in. tall. Flowers small with long hair-like stamens (pollen bearing male part of plant).

FOUND: Dry to moist soils in open areas along roads, on slopes and mountain ridges from about 6,000 to 8,500 ft. to above timberline throughout the West.

BLOOMS: June into early August.

USES AND FACTS: *Serecia* is from the Greek *Ser,* the Seres, an Indian people from whom the first silk came, or from the Latin *seris* meaning silky. Some species of *Phacelia* have been used by Indians as greens, particularly *P. racemosissima.*

SCORPIONWEED. *Phacelia hastata* (Waterleaf family: Hydro-phyllaceae).
SIZE: 6 to 15 in tall. Flowers small, in stalks shaped like a scorpion's tail.
FOUND: Dry, rocky places from 2,500 to 10,500 ft. from British Columbia throughout the Rocky Mountains and south to California.
BLOOMS: May to July.
USES AND FACTS: There are about 22 species of *Phacelia* in the Rocky Mountain area. Species of *Phacelia* are probably used to some extent by elk, deer and mountain goats.

118

WATERLEAF. *Hydrophyllum capitatum* (Waterleaf family: Hydrophyllaceae).

SIZE: 4 to 16 in. tall. Flowers in heads.

FOUND: Moist, rich soil, most often in shade from low valleys up to 9,000 ft. throughout the West.

BLOOMS: May and June.

EDIBLE: Young shoots and roots.

USES AND FACTS: When it rains, water is caught and held in the cavity of the leaf of this plant; hence, the name waterleaf. The young shoots are excellent in salad, and the roots may be cooked and eaten (Kirk, 1970).

119

VETCH; WILD PEA.   *Vicia cracca* (Pea family: Leguminosae)
SIZE: Plant is a vine 1½ to 3 ft. in length, prostate or twining around other vegetation.   Flowers ½ to ¾ in. long in spike-like racemes with flowers on one side of the stem.
FOUND: In fields, along roadsides and disturbed areas from low elevations to about 7,000 ft. throughout the West.
BLOOMS: June to August.
EDIBLE: Young stems and seedpods.
USES AND FACTS: This species of *Vicia* is from Eurasia and has escaped from domestic gardens and become widely naturalized.   All species of *Vicia* can be eaten.   The young stems and tender young seeds are supposed to be good when boiled or baked (Kirk, 1970).   This plant is also a favorite food of domestic sheep and deer, as well as other wild animals.   Makah Indian women soaked the roots and the water was used as a hair wash.   Squaxin Indians crush the leaves in bath water to take away soreness (Gunther, 1981).

LITERATURE CITED

Balls, E. K. 1962. *Early uses of California plants.* Univ. of Calif. Press, Berkeley and Los Angeles. 103 p.

Clarke, C. B. 1977. *Edible and useful plants of California.* Univ. of Calif. Press, Berkeley and Los Angeles 280 p.

Craighead, J. J., F. C. Craighead Jr. and R. J. Davis 1963. *A field guide to Rocky Mountain wildflowers.* Houghton Mifflin Co., Boston. 277 p.

Davis, R. J. 1952. *Flora of Idaho.* Brigham Young Univ. Press, Provo, UT. 836 p.

Elias, T. S. and P. A. Dykeman. 1982. *Field guide to North American edible wild plants.* Van Nostrand Reinhold Co., New York. 286 p.

Gunther, E. 1981. *Ethnobotany of Western Washington.* Univ. of Wash. Press, Seattle and London. 71 p.

Hardin, J. W. and J. M. Arena. 1974. *Human poisoning from native and cultivated plants.* Duke Univ. Press, Durham, N.C. 194 p.

Harrington, H. D. 1967. *Edible native plants of the Rocky Mountains.* Univ. of New Mexico Press, Albuquerque, N.M. 392 p.

Hart, J. and J. Moore. 1976. *Montana - native plants and early peoples.* The Montana Historical Soc., Helena, MT 75 p.

Hitchcock, C. L. and A. Cronquist. 1978. *Flora of the Pacific Northwest.* Univ. of Wash. Press, Seattle and London. 730 p.

Horn, E. L. 1972. *Wildflowers of the Cascades.* Touchstone Press, Beaverton, OR. 160 p.

Horn, E. L. 1980. *Wildflowers, the Pacific Coast.* Beautiful Amer. Publ. Co., Beaverton, OR. 144 p.

James, W. R. 1973. *Know your poisonous plants.* Naturegraph Publ., Healdsburg, CA. 99 p.

Kingsbury, J. M. 1964. *Poisonous plants of the United States and Canada.* Prentice-Hall, Inc., Englewood Cliffs, N.J.

Kinucan, K. W. and E. S. Kinucan. 1972. *Wildlife communities of Sun Valley: an ecological interpretation.* Sun Valley Creative Arts Center, Sun Valley, ID. 155 p.

Kirk, D. R. 1970. *Wild edible plants of the Western United States.* Naturegraph Publ., Healdsburg, CA. 326 p.

Lampe, K. F. and M. A. McCann. 1985. *AMA handbook of poisonous and injurious plants.* Amer. Medical Assoc., Chicago, IL. 432 p.

Munz, P. A. and D. D. Keck. 1959. *A California flora.* Univ. of Calif. Press, Berkeley and Los Angeles. 1681 p.

Niehaus, T. C. and C. L. Ripper. 1976. *A field guide to Pacific states wildflowers.* Houghton Mifflin Co., Boston. 432 p.

Spellenberg, R. 1979. *The Audubon Society field guide to North American wildflowers, Western region.* Chanticleer Press, Inc., New York. 862 p.

Spencer, E. R. 1968. *All about weeds.* Dover Publ., Inc., New York. 333 p.

Standley, P. C. 1943. *Edible plants of the Arctic region.* Bureau of Medicine and Surgery, Dept. of Navy, U.S. Govt. Printing Office, Washington, D.C.

Sweet, M. 1962. *Common edible and useful plants of the West.* Naturegraph Co., Healdsburg, CA. 66 p.

Taylor, R. J. and G. W. Douglas. 1975. *Mountain wild flowers of the Pacific Northwest.* Binford and Mort, Portland, OR. 176 p.

Taylor, R. J. and R. W. Valum. 1974. *Wildflowers 2. Sagebrush country.* The Touchstone Press, Beaverton, OR. 143 p.

Turner, N. 1975. *Food plants of British Columbia Indians. Part 1. Coastal Peoples.* Handbook #34, British Columbia Provincial Museum, Victoria. 264 p.

Urban, K. A. 1971. *Common plants of Craters of the Moon National Monument.* Craters of the Moon Natural History Association, Inc., Arco, ID. 30 p.

Winegar, D. 1982. *Desert wildflowers.* Beautiful Amer. Publ. Co., Beaverton, OR. 144 p.

INDEX

Sunflower family: Compositae	12, 13, 14, 15, 16, 42, 43, 44, 45, 46, 47, 48, 49, 50, 51, 72, 81, 82, 98, 99
Sweet cicely	57
Taraxacum officinale	45
Tea	23, 24, 32, 35, 58, 61, 62, 65, 78, 80, 84, 86, 95, 106, 113
Texas bluebonnet	106
Thalictrum fendleri	25
Thistle	82
Tiger lily	68
Tragopogon dubius	51
Trifolium sp.	78
Trillium ovatum	7
Twin flower	80
Umbelliferae	37, 38, 57, 58
Urticaceae	32
Urtica dioica	32
Valerian	39
Valerian family: Valerianaceae	39
Valeriana sp.	39
Vetch	120
Veratrum californicum	2
Verbascum thapsus	56
Vicia cracca	120
Violaceae	35, 61
Viola beckwithii	35
Viola purpurea	61
Violet	35, 61
Violet family: Violaceae	35, 61
Virgin's bower	26
Wake robin	7
Water cress	29
Water hemlock	37
Waterleaf	119
Waterleaf family: Hydrophyllaceae	36, 117, 118, 119
Wayside gromwell	67
Western groundsel	48
White bog orchid	9
White mule-ears	13
White wyethia	13
Wild flax	109
Wild iris	107
Wild lily of the valley	3

FIELD NOTES

FIELD NOTES

FIELD NOTES